Electricity from Wave and Tide

Electricity from Wave and Tide

An Introduction to Marine Energy

Paul A. Lynn, BSc (Eng), PhD
formerly
Imperial College London, UK

This edition first published 2014
© 2014 John Wiley & Sons Ltd

Registered office
John Wiley & Sons Ltd, The Atrium, Southern Gate, Chichester, West Sussex, PO19 8SQ, United Kingdom

For details of our global editorial offices, for customer services and for information about how to apply for permission to reuse the copyright material in this book please see our website at www.wiley.com.

Library of Congress Cataloging-in-Publication Data

Lynn, Paul A.
  Electricity from wave and tide : an introduction to marine energy / Paul A. Lynn.
      pages cm
  Includes bibliographical references and index.
  ISBN 978-1-118-34091-2 (hardback)
1. Tidal power-plants. 2. Ocean wave power. 3. Tidal power. I. Title.
  TK1081.L95 2014
  621.31′2134–dc23

                                                                                     2013019102

A catalogue record for this book is available from the British Library.

ISBN: 9781118340912

Typeset in 10/12pt Times New Roman by Laserwords Private Limited, Chennai, India

1 2014

Contents

Preface

The world's waves and tides are eternal and non-polluting, and the technology for converting their energy into grid electricity has reached an exciting stage. Many ingenious large-scale devices are currently being tested by developers as a prelude to commercialisation, in the confident hope that marine renewable energy will add significantly to conventional power generation in the coming decades.

This book introduces the history, theoretical background and practical development of today's wave and tidal stream devices to a wide readership including professionals, policy makers and employees in the energy sector needing an introduction or quick update. Its style and level also make it suitable background reading for university students and the growing number of thoughtful people who are interested in the contribution marine energy can make to 'keeping the lights on' in the twenty-first century. This is probably the first book to introduce wave and tidal stream technologies in a single volume and, although it assumes some basic familiarity with physics and maths, words are used every bit as much as symbols to give a descriptive flavour, enhanced by about 200 colour photographs and illustrations.

In more detail, Chapter 1 covers the historical background and Chapter 2 some of the key concepts underpinning today's practical developments. I have decided to devote Chapter 3 to electricity generation, for readers with little or no background in electrical engineering. Large-scale wave and tidal energy converters feed electricity into AC grid networks for the benefit of us all; yet electrical generation, grid connection and distribution are hardly ever explained in the context of marine energy and their terminology is mysterious to many people. I hope the account given here, which is very similar to that in my recent book on wind energy (also published by Wiley), will prove helpful.

Chapters 4 and 5 present case studies of modern wave and tidal stream devices, selected for their advanced state of development, including the testing of full-scale, or near full-scale, prototypes in sea conditions. I have relied heavily on the various developers for information about their devices, many of which have been, or are being, assessed at the internationally famous European Marine Energy Centre (EMEC) in Orkney, Scotland.

My interest in renewable energy goes back over 30 years, but I no longer have links with academia or industry and the selection and presentation of topics is my own. I claim no originality for the technical material, which has been gathered, sifted, and sorted from many websites, books, technical papers and articles. I see my role as hunter-gatherer, not master chef, and hope the menu will help advertise the remarkable developments currently taking place in the international quest for 'Electricity from Wave and Tide'.

Paul A. Lynn
Butcombe, Bristol, England
Summer 2013

Acknowledgements

I am very grateful to the following companies and organisations for information about their activities and devices, and for permission to use the excellent colour photographs and illustrations they have provided:

- Andritz Hydro Hammerfest, Hammerfest, Norway: Figures 5.1 and 5.2.
- Aquamarine Power Ltd, Edinburgh, Scotland: Figures 1.1, 2.15c, 4.6−4.12.
- Atlantis Resources Corporation, London, England: Figures 1.2, 2.33b, 2.38, 5.6−5.8.
- Datawell BV, Haarlem, The Netherlands: Figure 2.14.
- European Marine Energy Centre (EMEC), Orkney, Scotland: Figures 1.4, 2.43−2.45, 2.47−2.54.
- Marine Current Turbines Ltd, Bristol, England: Figures 2.33a, 5.9−5.13.
- Ocean Power Technologies Inc, New Jersey, USA: Figures 2.15b, 4.21, 4.23, 4.24.
- OpenHydro Ltd, Dublin, Ireland: Figures 5.14−5.19.
- Pelamis Wave Power Ltd, Edinburgh, Scotland: Figures 2.15a, 4.1−4.5.
- Pulse Tidal Ltd, Sheffield, England: Figures 5.20−5.25.
- Scotrenewables Tidal Power Ltd, Orkney, Scotland: Figures 2.41, 5.26−5.33.
- Tidal Generation Ltd, Bristol, England: Figures 2.33c, 5.34−5.37.

Acknowledgements

- Voith Hydro Ocean Current Technologies GmbH, Heidenheim, Germany: Figure 1.19.
- Voith Hydro Wavegen Ltd, Inverness, Scotland: Figures 2.15d, 4.13, 4.15, 4.16.
- Wave Dragon ApS, Copenhagen, Denmark: Figures 2.15e, 4.17–4.20.
- Wello Oy, Espoo, Finland: Figures 4.25–4.29.

The publishers acknowledge use of the above photographs and illustrations, which are reproduced by permission of the copyright holders.

I also acknowledge the following photos and illustrations obtained from Wikipedia: Figures 1.16–1.18, 2.21, 2.23, 2.24, 2.26–2.28, 2.35, 3.3.

The writer of an introductory book covering a wide field inevitably draws on many sources for information and inspiration. I make no claims for technical originality in the material presented and have tried to give an adequate list of references at the end of each chapter.

I would particularly like to thank staff at EMEC in Orkney for their enthusiastic cooperation and advice. I am also indebted to two books that have proved invaluable for clear explanations of difficult concepts which I have attempted to summarise. They are: *Ocean Wave Energy: Current Status and Future Perspectives* by Joao Cruz (editor), published by Springer in 2008; and *The Analysis of Tidal Stream Power* by Jack Hardisty, published by Wiley-Blackwell in 2009. Both are more comprehensive and advanced than my own offering, and I recommend them to anyone wishing to learn more about marine renewable energy.

Among the many figures in this book are 70 technical illustrations by David Thompson, who worked closely with me on two previous books, *Electricity from Sunlight* (Wiley, 2010), and *Onshore and Offshore Wind Energy* (Wiley, 2012). It has been a pleasure to repeat the collaboration.

Paul A. Lynn

1 Introduction

1.1 Marine energy and Planet Earth

For over a century most of the electricity used in our homes, offices and factories has been generated in large power plants based on fossil fuels and, in some countries, nuclear reactors and hydroelectric turbines. But as the new millennium gets into its stride important changes are taking place in the how, where and why of electricity generation due to increasing concerns about climate change, fossil fuel depletion and the risks of nuclear power. Terms such as *renewable, sustainable* and *carbon-free* have entered the popular imagination and most experts and politicians now accept that a major redirection of energy policy is essential if Planet Earth is to survive the twenty-first century in reasonable shape.

For the last few hundred years humans have been using up fossil fuels that nature took around 400 million years to form and store underground. A huge effort is now under way to develop energy systems that make use of natural energy flows in the environment – including those produced by waves and tidal streams. This is not simply a matter of fuel reserves, for it is becoming clearer by the day that, even if those reserves were unlimited, we could not continue to burn them with impunity. Today's scientific consensus assures us that the resulting carbon dioxide emissions would very likely lead to a major environmental crisis. So the danger is now seen as a double-edged sword: on the one side, fossil fuel depletion; on the other, the increasing inability of the natural world to absorb emissions caused by the burning of what fuel remains, leading to accelerated global warming.

Electricity from Wave and Tide: An Introduction to Marine Energy, First Edition. Paul A. Lynn.
© 2014 John Wiley & Sons, Ltd. Published 2014 by John Wiley & Sons, Ltd.

Things were not always like this. Back in the 1970s there was little public discussion about energy sources and engineering courses in universities paid little attention to them. The environmental movement was in its infancy, far removed from the mainstream political agenda, and its proponents were often dismissed as eccentric busybodies. Few people had any idea how the electricity they took for granted was produced, or that the burning of coal, oil and gas might be building up global environmental problems. Those who were aware tended to assume that the advent of nuclear power would prove a panacea, a few even claiming that nuclear electricity would be so cheap that it would not be worth metering!

Yet even in those years a few brave voices suggested that all was not well. In his famous book *Small is Beautiful* [1], first published in 1973, E.F. Schumacher poured scorn on the idea that the problems of production in the industrialised world had been solved. Modern society, he claimed, does not experience itself as part of nature, but as an outside force seeking to dominate and conquer it. And it is the illusion of unlimited powers deriving from the undoubted successes of much of modern technology that is the root cause of our present difficulties, in particular because we are failing to distinguish between capital and income components of the earth's resources. We use up capital, including coal, oil and gas reserves, as if they were steady and sustainable income, but they are actually once-and-only capital. Schumacher's heartfelt plea encouraged us to start basing industrial and energy policy on what we now call sustainability, recognising the distinction between capital and income and the paramount need to respect the planet's finite ability to absorb the polluting products of industrial processes – including electricity production.

Schumacher's message, once ignored or derided by the majority, is now seen as mainstream. For the good of Planet Earth and future generations we have started to distinguish between capital and income and to invest heavily in renewable technologies that produce electricity free of carbon emissions. In recent years the message has been powerfully reinforced by former US Vice President Al Gore, whose inspirational lecture tours and video presentation *An Inconvenient Truth* [2] have been watched by many millions of people around the world.

Into this melting pot of hopes and concerns fall a number of promising renewable technologies based on the immense natural energy flows in Planet Earth's environment. These include winds and the ocean waves they produce (see Figure 1.1), tides and tidal streams (see Figure 1.2), and sunlight falling on the Earth's surface. All are eternal and inexhaustible; nothing is 'wasted' if we ignore them because they are there anyway. They are income, not capital, and we should surely regard them as precious gifts of nature to be harnessed in ways that are technically efficient, economic and environmentally

Figure 1.1 Harnessing wave energy (Aquamarine Power Ltd).

sensitive. All this represents a hugely challenging and inspiring agenda for engineers and scientists – now and for the rest of the century.

Perhaps we should consider the meaning of renewable energy a little more carefully. It implies energy that is sustainable in the sense of being available in the long term without significantly depleting the Earth's capital resources, or causing environmental damage that cannot readily be repaired by nature itself. In his excellent book *A Solar Manifesto* [3], German politician Hermann Scheer considered Planet Earth in its totality as an energy conversion system. He noted how, in its early stages, human society was itself the most efficient energy converter, using food to produce muscle power and later enhancing this with simple mechanical tools. Subsequent stages – releasing relatively large amounts of energy by burning wood; focussing energy where it was needed by building sailing ships for transport and windmills to grind grain and pump water – were still essentially renewable activities in the above sense.

What really changed things was the nineteenth century development of the steam engine for factory production and steam navigation. Here, almost at a stroke, the heat energy locked in coal was converted into powerful and highly concentrated motion. The industrial society was born and ever since we have continued burning coal, oil and gas in ways which pay no attention to the natural rhythms of the earth and its ability to absorb wastes and by-products, or to keep providing energy capital. Our approach has become the opposite of renewable and it is high time to change priorities.

Figure 1.2 Transporting a tidal stream turbine (Atlantis Resources; Mike Roper (photographer)).

Since the reduction of carbon emissions is a principal advantage of wave, tidal and other renewable technologies, we should recognise that this benefit is also proclaimed by supporters of nuclear power. But frankly they make strange bedfellows, in spite of sometimes being lumped together as 'carbon-free'. It is true that all offer electricity generation without substantial carbon emissions, but in almost every other respect they are poles apart. The renewables, including wave and tidal stream energy, give us the option of widespread, relatively small-scale electricity generation, but nuclear must, by its very nature, continue the practice of building huge centralised power stations. Waves and tides give us 'free fuel' and produce no waste in operation; the nuclear industry is beset by problems of radioactive waste disposal. On the whole renewable technologies pose no serious problems of safety or susceptibility to terrorist attack – advantages which nuclear power can hardly claim. Finally, there is the issue of nuclear proliferation and the difficulty of isolating civil nuclear power from nuclear weapons production. Taken together these factors amount to a profound divergence of technological expertise and political attitudes, even of philosophy. It is not surprising that most environmentalists are unhappy with the continued development and spread of nuclear power, even though some accept that it is proving hard to avoid. In part, of course, they claim that this is the result of policy failures to invest sufficiently in the benign alternatives over the past 30 or 40 years.

However, we must be careful not to assume that renewable energy is an easy answer. For a start it is generally diffuse and intermittent. Quite often, it is unpredictable. The design and manufacture of efficient machines to harness natural energy flows pose big technical problems, and although the 'fuel' may be free and the waste products minimal, up-front investment costs tend to be large. There are certainly major challenges to be faced and overcome as we develop a new energy mix for the twenty-first century.

Our story now moves on to modern wave and tidal stream technology, currently enjoying rapid progress and poised to make a significant contribution to electricity generation in the coming decades. But before getting involved in the details, we should consider the natural resources that promise to help wean us away from our addiction to fossil fuels.

1.2 Marine resources

1.2.1 Waves of the world

Surface waves on the world's oceans are generated by the wind. They are not formed instantly but build up over time and with distance, known as the *fetch*. Waves produced by a storm, arriving from afar over deep water, produce a regular *swell* which may take hours or days to form and travel hundreds or even thousands of kilometres across an ocean with very little loss of energy (see Figure 1.3). But as waves approach the shore and move

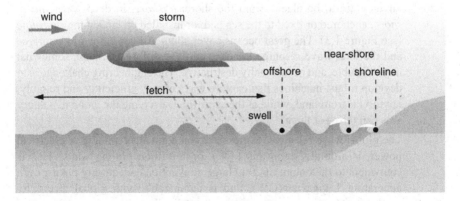

Figure 1.3 Wind-generated ocean waves.

Figure 1.4 Waves approaching a shoreline (EMEC).

into shallow water, they slow down, increase in height and start to break, dissipating lots of energy (see Figure 1.4). Wave characteristics close to shore can be very different from those of a regular, deep-water, swell.

From the engineer's point of view wind-generated waves represent a valuable source of renewable energy for generating electricity using *wave energy converters* [4, 5]. The design of effective machines depends on how far they are to be placed from the shore: *offshore*, in deep water; *near-shore*, anchored or fixed to the sea bed; or installed on land at the *shoreline* (see Figure 1.3). The great oceans cover about 70% of the world's surface and the total wave resource is huge; but it is diffuse, variable, somewhat unpredictable, and occasionally destructive. The engineering challenge is to develop robust machines that capture wave energy efficiently and reliably, not too far from land, while at the same time surviving the worst that angry seas can throw at them.

The global wave resource, expressed as an equivalent amount of electrical power, is around 2 terawatts (TW), or 2 million million watts. This is equivalent to the output of 2000 large conventional electricity plants, each generating 1 gigawatt (GW), and is comparable with global electricity production. However, wave resources are distributed very unevenly across the world's oceans and countries with strong prevailing winds and exposed coastlines are the most favoured. A good example is the UK's coastal waters which receive, on average, wave power roughly equivalent to the nation's electricity demand. Although the exploitable resource in terms of practicality and economics is only a small percentage of the total, there is no doubt that ocean waves could make a significant contribution to an energy mix based increasingly on renewables.

Why do some maritime nations receive much more wave energy than others? The answer to this question is closely related to the world's major wind patterns, set up as the earth spins on its axis. Variations in atmospheric pressure caused by differential solar heating propel air from high pressure to low pressure regions, generating winds that are greatly affected by the earth's rotation and tend to occupy certain latitudes.

The investigation of *latitudinal wind belts* has a long history. For centuries the captains of sailing ships depended on reliable north-east and south-east *trade winds* to speed them on their way, and tried to avoid the *horse latitudes* that could becalm them. They also had to contend with strong but variable *westerlies* that blow in the mid-latitudes between about 40° and 60°, north and south (see Figure 1.5). It is hardly surprising that wind meteorology exercised some famous minds throughout the great age of sail. Edmond Halley (1656–1742), an English astronomer best known for computing the orbit of *Halley's comet*, published his ideas on the formation of trade winds in 1686, following an astronomical expedition to the island of St Helena in the South Atlantic. The atmospheric mechanism proposed by George Hadley (1685–1768), a lawyer who dabbled productively in meteorology, attempted to include the effects of the Earth's rotation – a theory that was

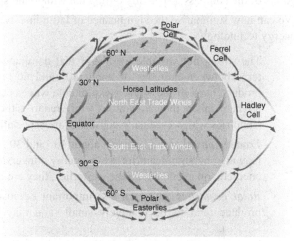

Figure 1.5 Atmospheric cells and latitudinal wind belts.

subsequently corrected and refined by American meteorologist William Ferrel (1817–1891).

The contributions of Hadley and Ferrel to our understanding of latitudinal wind belts, and the waves they generate, are acknowledged in the names given to the atmospheric 'cells' shown in Figure 1.5. Essentially these are produced by the steady reduction in solar radiation from the equator to the poles. The associated winds, rather than flowing northwards or southwards as we might expect, deflect to the east or west in line with the *Coriolis effect*, named after French engineer Gaspard Coriolis (1792–1843), who showed that a mass (in this case, of air) moving in a rotating system (the Earth) experiences a force acting perpendicular to both the direction of motion and the axis of rotation.

The *Hadley cells*, closed loops of air circulation, begin near the equator as warm air is lifted and carried towards the poles. At around 30° latitude, north and south, they descend as cool air and return to complete the loop, producing the *north-east* and *south-east trade winds*. A similar mechanism produces *polar cells* in the arctic and antarctic regions, giving rise to *polar easterlies*.

The *Ferrel cells* of the mid-latitudes, sandwiched between the Hadley and polar cells, are less well defined and far less stable. Meandering high-level *jet streams* tend to form at their boundaries with the Hadley cells, generating localised passing weather systems. This makes the coastal wind patterns – and ocean climates – of countries such as Norway, Denmark, Britain and Ireland strong but famously variable. So although the prevailing winds are *westerlies*, they are quite often displaced by flows from other points of the compass, especially during the winter and spring months.

We can now summarise the significance of latitudinal wind belts for wave energy technology:

- The strong but variable *westerlies* that dominate global wind patterns between latitudes of about 40° and 60° (north and south) produce most of the world's exploitable wave energy. In the southern hemisphere they are famously referred to as the *Roaring Forties*. Countries with west-facing coastlines are especially favoured.

- *Trade winds* blowing between about 10° and 30° (north and south) may also be significant for wave energy conversion. Although less energetic on average than the westerlies, they are more consistent.

- *Polar easterlies* are much less important because the swells they produce tend to be relatively small (and may be hampered by sea ice).

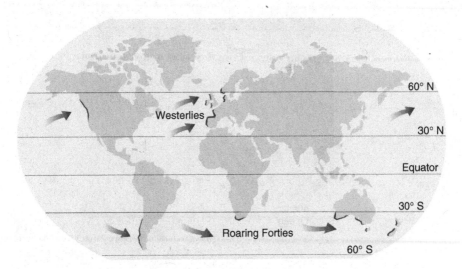

Figure 1.6 Coastlines with large wave energy resources.

The dominance of wave energy produced by prevailing westerly winds is emphasised in Figure 1.6, which shows coastlines that receive heavy swells generated over long fetches of ocean. Not surprisingly, they lie in countries presently showing great interest in wave energy, principally:

- **In Europe:** UK, Norway, Denmark, Ireland, France, Spain and Portugal.
- **In North America:** USA and Canada.
- **In the Southern Hemisphere:** Australia, New Zealand, Chile and South Africa.

We now come to a very important question: how much power do ocean waves possess as they travel across an ocean and approach a coastline? The first point to make is that it depends on the distance from the shoreline. Wave power is greatest well offshore in deep water but, as the waves move into shallower water, friction with the sea bed and their tendency to break cause energy losses. The usual way of expressing power levels is in terms of average kilowatts per metre length of wave front ($kW\,m^{-1}$). Figure 1.7 shows typical values well off the coastlines of Western Europe favoured with some of the world's best wave resources. We see that, for example, the average power of waves approaching the west coast of Portugal is around 50–60 $kW\,m^{-1}$; off the west coast of Scotland, one of the most productive areas in the world, around 70 $kW\,m^{-1}$; and along the coast of Norway, around 50 $kW\,m^{-1}$ and diminishing steadily towards the Arctic Circle.

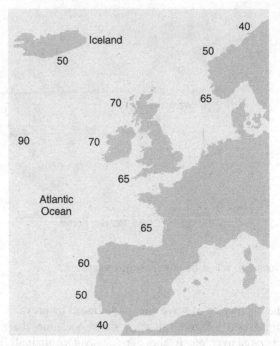

Figure 1.7 Average values of wave power off the coasts of Western Europe, expressed in kilowatts per metre of wave front.

Out in the Atlantic ocean it can reach 90 kW m^{-1}. Elsewhere in the world such values are only matched along certain coastlines in Australia, New Zealand and Chile, especially those facing the *Roaring Forties* that blow unhindered from west to east across the Southern Ocean at latitudes around 40°S (see Figure 1.6). It is hardly surprising that the most powerful wave climate in the world, averaging some 140 kW m^{-1}, is found at latitude 48°S in the Southern Ocean – but so far from civilisation that it is extremely unlikely ever to be harnessed!

Such values are certainly impressive. It is sometimes said that 25 kW m^{-1} is enough to interest wave energy enthusiasts, but the 50–70 kW m^{-1} found off the coastlines of Western Europe is clearly a great deal better and represents a *power concentration* rarely found in natural energy flows. For example, it is far greater than that of the airstreams harnessed by wind turbines. Essentially this is because wave power is built up and concentrated gradually over long stretches of ocean; and because seawater is far denser than air – a point we will return to in the next chapter.

However, power concentration is not the only criterion of interest to designers of wave energy converters. It is certainly important because the more power a machine of given size can capture the better but *variability* is also a major issue. It is all very well to choose a location with a high average wave power, promising high annual energy capture, but it is likely to produce occasional peaks that place great mechanical stresses on wave machines and may even threaten their survival. *Reliability* and *survivability* are crucial to economic justification and designers sleep better at night if their devices are located in somewhat calmer, less variable, waters. This makes ocean waves off the coastlines of countries such as Japan, Peru and Ecuador potential candidates for wave energy conversion even though average power concentrations rarely exceed $30\,\mathrm{kW\,m^{-1}}$.

The variability of wave power at a particular location occurs over widely differing time scales:

- *Short term.* Successive waves in an ocean swell are not all equal in size, but vary in a somewhat random fashion. A wave energy converter must cope with variable power levels over time scales from seconds to minutes, even when the sea state is nominally steady. In heavy seas there may be occasional 'rogue' waves. The situation close to shore, where waves start to break, is even less predictable.

- *Medium term.* Ocean swells caused by storms at sea build up and decay over time scales from hours to days. Such variability is especially strong in the mid-latitudes, for example off the western coasts of the UK and Ireland.

- *Long term.* The average wave power in most locations varies substantially according to the season of the year. In the mid-latitudes of the northern hemisphere this *seasonality* may easily result in a $3:1$ ratio between wave resources in the winter and summer months. In the southern hemisphere, seasonal variations tend to be considerably smaller.

Wind-generated waves are an important energy resource but their variability presents major challenges to designers and engineers. In Chapter 4 we will meet a number of wave energy converters that illustrate current approaches to harnessing this powerful but unruly gift of nature.

1.2.2 Tides of the world

The main influences on ocean tides are the gravitational attractions of the moon and sun, and the earth's rotation. If you stand on a seashore you will likely see the water rising and falling twice a day or, more precisely, twice every 24 hours, 50 minutes and 28 seconds, the moon's apparent period of rotation about the earth. Such tides are referred to as *semi-diurnal*. In some

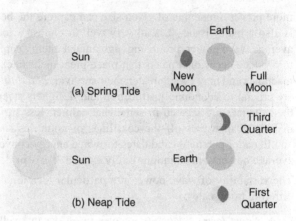

Figure 1.8 Relative positions of sun, moon and earth giving rise to spring and neap tides.

locations there is only one high and one low tide each day, referred to as *diurnal*.

The highest and lowest points reached by a tide are known as *high water* and *low water*, and the vertical difference between them as the *tidal range*. At a given location on an ocean shore the tidal range varies over the course of each month according to the relative positions of sun, moon and earth. The sun's attraction is only about half as great as the moon's because, although the sun is massive, it is much further away than the moon and gravitation is governed by an inverse-square law of distance. When the sun and moon are in line (see Figure 1.8), giving a *full moon* or a *new moon*, their gravitational attractions act together and the tidal range is greatest. This is referred to as a *spring tide*. But when the sun and moon are at right angles, the moon is said to be in its *first* or *third quarter* and the tidal range is a minimum. This is known as a *neap tide*. In most locations the dominant cyclic variation in tidal range repeats twice a month.

The moon's orbit around the earth is slightly elliptical rather than circular, and the earth is not at the centre of the ellipse. This means that there is a time in each month when the moon is nearest to the earth (*perigee*), and another when it is furthest away (*apogee*). In a few locations tidal ranges are more influenced by this effect than by the more familiar spring-to-neap variations, and the resulting tides are referred to as *anomalistic*.

A further factor affecting tidal range is the moon's *declination* – its angular offset with respect to the earth's equatorial plane – which tends to make the two tides of a day unequal in range. When this effect is especially pronounced the tides are referred to as *declinational*.

We see that subtleties in the moon's orbit affect tidal ranges in a number of ways. The situation is further complicated by the influence of coastal geography and seabed geometry (*bathymetry*) as tidal undulations or 'bulges' make their daily journey around the earth's surface. In the open ocean tidal ranges are less than 1 m, but close to land they can be greatly increased by the way huge volumes of water work their way round continents, build up against coastlines, force their way through channels, or enter bays and estuaries. Tidal ranges in Canada's Bay of Fundy can reach 17 m, and in England's Bristol Channel 14 m, but in parts of the Mediterranean, Baltic and Caribbean seas they are close to zero. Coastal geography also affects the temporal patterns of tides, which may depart dramatically from those experienced in deep water offshore. But whatever the details, the tides in a particular location go through a monthly cycle.

So far we have discussed tidal ranges, the regular 'ups and downs' of sea level which, in a few locations, are used to generate electricity with *tidal barrages* [6]. However our focus in this book is on *tidal streams* [7], the oscillating horizontal currents that accompany the rise and fall of tides. Tidal stream technology captures moving water's *kinetic* energy, whereas tidal barrages make use of stored water's *potential* energy. The global tidal stream resource is comparable to that of wave energy (say 2–3 TW), of which perhaps 3% is reasonably accessible for electricity generation. But, as we shall see, it is very unevenly distributed.

Tidal power is more reliable than wave power because it depends on highly predictable movements of the earth, moon and sun. This makes the technology attractive to energy planners who like to know well in advance how much electricity they will be offered, and when it will arrive! Another advantage of tidal streams is their comparative docility – although we must be careful to emphasise the word *comparative* because fast streams are generally turbulent and can place very high stresses on turbines and other devices. The marine environment is invariably tough, at and below the sea surface.

Tidal streams are reasonably predictable, but they are certainly not constant. A stream consists of a *flow* phase as the tide rises, alternating with an *ebb* phase as it falls, and a good stream for energy generation achieves high peak velocities in both phases. The times when the current ceases are referred to as *slack water*, and in most locations they occur close to high and low water. Efficient devices must generate electricity on both ebb and flow, in other words they are *bi-directional*, unlike land-based *run-of-river* turbines which extract energy from water flowing in one direction only.

Figure 1.9 shows the flow pattern of a strong tidal stream over an 11-day period encompassing spring and neap tides, in a location where the tides exhibit a straightforward semi-diurnal pattern (two similar

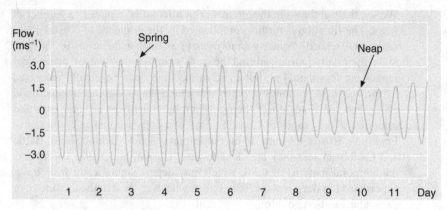

Figure 1.9 Flow rates of a tidal stream.

tides per day). At spring tides the peak flow rate exceeds 3 m s^{-1} in both directions (positive for the flow phase, negative for the ebb phase), but at neap tides it falls to about 1.5 m s^{-1}. In many locations the basic pattern is modified by turbulence, which can affect peak flows considerably. Also, there are variations with water depth; flow rates are greatest near the surface and reduce as the seabed is approached.

In this book we generally quote tidal flow rates in metres per second. However alternative units such as kilometres per hour, miles per hour and knots (nautical miles per hour) are sometimes used – the latter especially by sailors and mariners. The relationship between them is:

$$1 \text{ m s}^{-1} = 3.60 \text{ kph} = 2.24 \text{ mph} = 1.95 \text{ knots}.$$

For convenience it is helpful to remember that the speed in knots is very close to twice its value in metres per second.

The flow pattern of a tidal stream correlates closely with the rise and fall of local tides, although their influence may be hard to analyse. Many tides produce far more complex flow patterns than the straightforward spring–neap variation shown in Figure 1.9. The flow volume along a curved or irregular channel remains constant, even though its depth, speed and direction may vary continuously. The formation of large eddies depends on the channel's shape upstream, not downstream. Generally speaking, flow magnitudes vary over the lunar cycle, peaking a few days after each new or full moon and increasing still further around the time of the *equinoxes* in March and September. All these effects are very significant for the output of a tidal turbine, which oscillates in sympathy with the complex flow pattern.

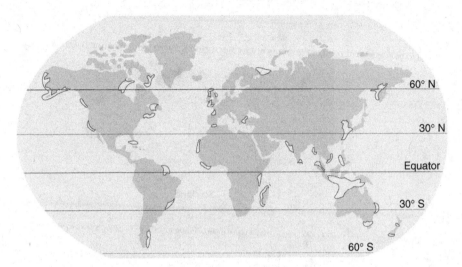

Figure 1.10 Coastal regions of the world with good tidal stream resources.

It is important to realise that strong tidal streams do not necessarily accompany large tidal ranges. For example you might be standing on a shore watching the tide rise and fall through many metres, yet see no horizontal currents strong enough to work a turbine. Conversely, a relatively small tidal range may produce a vigorous tidal stream through a narrow channel between an island and nearby coastline, or through the inlet of a large bay or estuary. The relationship between tidal ranges and tidal streams depends on complex interactions between the movement of large volumes of water around continents and islands, and local effects including coastal features and the shape of the seabed (bathymetry). In addition, when sea enters a bay or estuary the tidal patterns may be greatly influenced by its depth, length and area [6].

So where does all this complexity lead us in terms of the regions of the world which have most potential for tidal stream technology? Figure 1.10 summarises the global scene, and it is interesting to compare it with Figure 1.6 which showed the distribution of wave energy potential. For a start, the powerful wave resources off the west-facing coasts of Portugal, Chile and Australia are not matched by comparable tidal stream resources. Conversely, ocean areas to the north of Australia and east of China with little wave energy potential have some very powerful tidal streams. With few exceptions, powerful wave and tidal stream resources do not coincide.

One of the few nations in the world to be doubly blessed is the UK, and especially Scotland, whose waves and tidal streams, backed up by technical

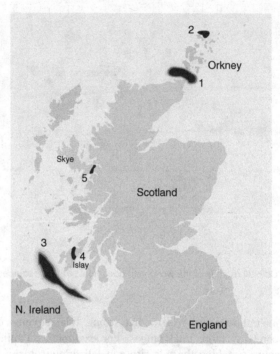

Figure 1.11 Important tidal stream locations in Scottish waters.

innovation and political will, are producing world leadership in marine renewable energy. We may use some of Scotland's prime tidal stream locations for illustrative purposes. Figure 1.11 shows three areas of coastal water and two narrow channels formed by the complex geography of the west coast and inner Hebridean islands:

1. *Pentland Firth.* Often considered Europe's potential 'powerhouse' of marine renewable energy, this large area of water between the north coast of Scotland and the Orkney Islands produces some of the best tidal streams in the world and boasts about 25% of the total European resource. Peak flow rates up to $4 \, \text{m s}^{-1}$ at spring tides are routinely recorded. It seems possible that thousands of turbines will be placed in the Pentland Firth over the next 30 years, with a total installed capacity up to 2 or 3 GW.

2. *Orkney Islands.* Some powerful tidal streams exist around these scattered islands, especially *Westray* and *Eday*. The resource is about a fifth of that in the Pentland Firth. It is no coincidence that the European Marine Energy Centre (EMEC), the international leader in its field, is based in Orkney (see Section 2.5.2).

3. *Isle of Islay.* A large area of water to the west and southwest of Islay, separating Scotland from Northern Ireland, shows great tidal stream potential. The total resource is comparable with that of the Pentland Firth. Peak flows at spring tides are typically 2.5 m s^{-1} and major installations of tidal turbines are confidently expected in the next 30 years.

4. *Sound of Islay.* This narrow channel separating the islands of Islay and Jura is about 10 km long and 1 km wide at its narrowest point (see Figure 1.12). The total resource is far smaller than those mentioned above but the location offers vigorous tidal streams, proximity to shore, good shelter from Atlantic storms and high waves, and a nearby electricity grid – an ideal situation for an array of tidal stream turbines (see Section 5.2.1).

5. *Kyle Rhea, Isle of Skye.* Another narrow channel with fast tidal streams, Kyle Rhea separates the Isle of Skye from the mainland a few kilometres south of Skye Bridge. Approximately 3 km long by 1 km wide, its tidal resource is smaller than the Sound of Islay, but the relatively shallow water may suit tidal stream machines of the 'oscillating hydrofoil' variety (see Section 5.2.5), as well as more conventional turbines.

Figure 1.12 A fast-flowing tidal stream in the Sound of Islay (Paul A. Lynn).

These five sites represent a good selection for testing and proving today's tidal stream machines. Relatively small, sheltered, locations such as the Sound of Islay and Kyle Rhea are ideal for installing and proving the viability of prototype arrays. Operational experience gained from such sites will give developers confidence to move into larger, more exposed, sea areas including the Pentland Firth where the dream of making a significant contribution to national electricity supplies can become a reality.

1.3 A piece of history

1.3.1 Working with waves

The history of wave energy conversion goes back over 200 years and may conveniently be divided into two phases. The first ended with the 1973 'oil shock', when oil-producing nations in the Middle East showed their determination to exercise greater control over the price and availability of their 'black gold'. This acted as a wake-up call for western governments to consider alternative energy sources, including ocean waves. The second phase, from 1973 to the present day, is very much the modern history of wave energy, with its moments of optimism and setback, culminating in sustained interest and commitment by governments to support a fledgling industry moving towards commercialisation.

The first patent covering the design of a wave machine was granted to a Monsieur Girard and his son in Paris in 1799. Their idea was startling: to dock naval warships and use their bobbing up and down on the waves to rock long wooden beams. The heaving motion of the beams, acting as levers with their fulcrums on the shore, would generate mechanical power to drive pumps, saws and other machinery. In a moment of optimism the patent declared that 'with a vessel suspended at the extremity of a lever, one may conceive the idea of the most powerful machine that has ever existed'. However the Girard plan was never realised. Perhaps it was considered too outlandish; or maybe the naval authorities in France had more pressing duties for their warships. After all, the year before had seen a heavy defeat of Napoleon's navy at the Battle of the Nile, and six years later came the battle of Trafalgar.

The nineteenth century spawned many new proposals for transmitting the oscillating motion of waves to pumps and other machinery using various types of mechanical linkage. By 1973 well over 1000 wave energy patents had been registered in Western Europe, North America and Japan, including 340 in the UK. Among various early efforts to translate paper designs into

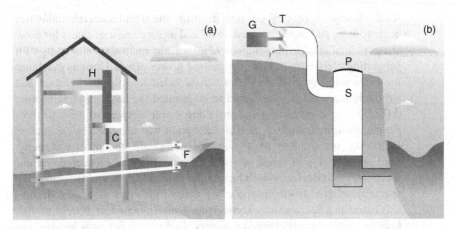

Figure 1.13 Two early examples of wave machines: (a) P. Wright's 'wave motor' of 1808 and (b) the Bochaux-Praceique 'oscillating water column' of 1910.

practical machines, we will consider two that are widely considered of historical importance [4]. Their operating principles are illustrated in Figure 1.13.

The 'wave motor' shoreline system invented by P. Wright was patented in the USA in 1898. Essentially it consisted of one or more hinged floats (F) which rode the approaching waves. Each float transmitted vertical motion to a connecting rod (C) operating a hydraulic pump (H) that could be used to power a wide variety of machinery. This was one of many devices proposed for the wave-rich beaches of Southern California in the 1890s. Only a few designs made it to full-scale, and only the Wright example still exists – buried, presumably in a very sad condition, beneath the sand of Manhattan Beach.

Among European efforts at working with waves, the system built by Monsieur Bochaux-Praceique at Royan, near Bordeaux, is historically important both for its overall design and because it successfully generated up to 1 kW of electricity to power and light his home (Figure 1.13b). A shaft (S), sunk in a nearby cliff, was sealed at the top by a pressure cap (P) and connected to the sea by a short tunnel below low water level. Waves caused the water level in the shaft to oscillate, producing pressure fluctuations in the air column above and driving a turbine (T) connected to a generator (G). This was an early example of an *oscillating water column* – a very different approach to wave energy conversion from that taken by Wright and most other early inventors. By today's standards it was, of course, hopelessly uneconomic; just imagine carrying out major earthworks in order to generate a mere 1 kW of intermittent electricity!

As the twentieth century got into its stride the world's energy industries became more and more focussed on oil and its supreme usefulness for powering internal combustion engines. Coal was the main source of energy for generating electricity in large centralised power plants. Curious machines for extracting power from ocean waves faded into the background and although many patents continued to be granted the costs and engineering difficulties of constructing and installing viable devices seemed, to most people, insurmountable. A notable exception was provided by navigation lights for buoys at sea which need only a small amount of electricity. Starting in 1945, Japanese inventor Yoshio Masuda pioneered various wave-activated devices which he mounted and tested on an 80 m long ship specially adapted for the purpose [4]. One of his most successful designs was based on a float and long vertical tube which pumped air through a small turbine-generator and charged a battery. From 1965 onwards hundreds of Masuda devices provided electricity for navigation lights at sea. However, such a low-power application hardly addressed the question whether large wave energy converters could generate substantial amounts of electricity for use onshore.

As already noted, the first 'oil shock' of 1973 acted as a turning point. In the USA, President Carter gave serious support to renewable energy technologies, principally wind and solar power, and although President Reagan's subsequent administration proved far less enthusiastic, the die had been cast and international attention was now focussed on the increasing price and eventual depletion of the world's oil supplies. It is only fair to add that large-scale wave machines had few supporters at this stage – at least, not until the publication in 1974 of an article [8] in the scientific journal *Nature* by Professor Stephen Salter of the University of Edinburgh.

It would be hard to overestimate the effect of the Salter article on wave energy research and development, both in the UK where it was born, and internationally. Here was a talented engineer and researcher who re-examined wave energy conversion from first principles, discarding the widespread assumption that waves are only effective at generating up-and-down motion, and able to back up his theoretical insights with a splendid series of practical experiments [9]. And so the *Edinburgh Duck*, popularly known as the *Nodding Duck*, was born. Although it subsequently progressed as far as a 1/10th scale model, its early promise was never realised, partly because politicians and energy policymakers tended to slip temporarily back into denial about the oil problem, and partly – as many would claim – because the UK government under Margaret Thatcher had a love of nuclear power and an irrational dislike of anything renewable. In any case, funding for the UK's wave energy programme was severely cut in the late 1970s and the Nodding Duck was a principal casualty. Yet it has had a

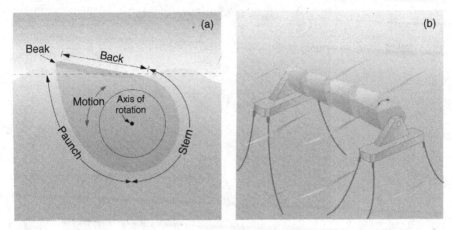

Figure 1.14 Nodding Ducks: (a) a single device and (b) an array.

profound and lasting influence on the wave energy community, stimulating innovation in design and an appreciation of the challenges ahead.

The Nodding Duck is illustrated in Figure 1.14. Essentially, it consists of a device shaped like a large cam, able to 'nod' backwards and forwards about a horizontal axis under the action of incoming waves. Its oscillatory motion is converted into electricity by a hydraulic-electric subsystem. The highly original design concept is based on the way water particles in an ocean swell actually move as waves approach the device – not just up and down as might be expected, but in a circular path. We shall have more to say about this in Chapter 2. Nodding is caused partly by the dynamic pressure of moving water on the Duck's 'paunch', and partly by changing hydrostatic pressure on its buoyant 'beak'. Such devices can achieve remarkable efficiencies, up to 90%, in converting wave energy to mechanical energy. To generate substantial amounts of power, an array of nodding ducks can be mounted on a common shaft set parallel to the incoming wave front.

The surge of interest in wave energy in the 1970s generated other ideas [5] that influence designers and engineers to this day, including:

- *Contouring raft [10].* British engineer Christopher Cockerell, best known for his invention of the Hovercraft, was also active in wave energy research and suggested using a set of hinged rafts to follow the ups and downs of waves as they passed (see Figure 1.15). The relative motion between each pair of rafts at the hinges would supply power to a hydraulic subsystem. In 1978 his company tested a 1/10th scale, 3-raft model in waters off the Isle of Wight in southern England

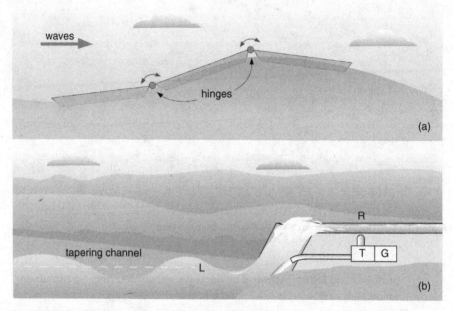

waves

hinges

(a)

R

tapering channel

T | G

L

(b)

Figure 1.15 Tapping wave energy using (a) contouring rafts and (b) a tapering channel.

and, for a time, the contouring raft was seen as a strong competitor of the nodding duck.

- *Wave focussing [4].* This term is used to describe various techniques for concentrating the energy of long wave fronts onto relatively small power conversion devices. They include the *antenna effect* pioneered by Johannes Falnes and colleagues at the University of Trondheim in Norway who have made major contributions to wave energy research and development over many years [11]. Successful focussing of surging waves in a narrow tapering channel was achieved by the Norwegian *Tapchan* system installed in 1985. The action of such a system is illustrated in Figure 1.15b. Incoming waves increase in height as they move up the channel, finally 'overtopping' the lip of a man-made or natural reservoir (R) set a few metres above mean sea level (L). This converts the kinetic energy of the waves into potential energy that drives a turbine (T) and generator (G).

We have now introduced some of the most influential devices and systems in the 200-year history of wave energy conversion: mechanical 'wave motors'; oscillating water columns; floating buoys; nodding ducks; contouring rafts and tapering channels. Their operating principles differ widely and it has never been clear which will prove most effective and economical for use in different wave climates offshore, near-shore and onshore – or whether entirely new concepts and designs will take over. As we shall discover in Chapter 4, this question is still very much alive today.

1.3.2 Tapping tides

The science of ocean tides – understanding their origins and the factors affecting their rhythms and ranges – has a long history. As we shall see in Section 2.3.2, famous names including Newton, Laplace and Kelvin contributed over the centuries, and today's tidal science is so highly advanced that the fluctuating times and patterns of tides may generally be predicted with a high degree of accuracy [7].

The technology used to tap tidal energy also goes back a long way. We will start with a brief account of systems which use the *potential* energy of stored water, then move on to the main focus of this book – systems that tap the *kinetic* energy of tidal streams.

Tidal mills, which store seawater in a millpond or *tidal basin* on the flood tide and generate mechanical power by releasing it on the ebb, were developed along the Atlantic coasts of Europe from the Middle Ages onwards. Typically, a dam with sluice gates traps water in a tidal basin, subsequently releasing it back into the sea via a large waterwheel (see Figure 1.16). Essentially, tidal mills are a variation on more conventional water mills worked by streams or rivers, with the added complication that power generation is tied, day and night, to a timetable set by the tides. A few tidal mills continued working well into the twentieth century but the advent of electricity grids fed by large power plants rendered them obsolescent. If tidal power was to be tapped in the future, it would have to generate substantial amounts of electricity rather than a modest amount of mechanical power for strictly local use.

In modern times, the largest and best-known tidal energy project is the La Rance tidal barrage on the Brittany coast of France, shown in Figure 1.17. Commissioned in 1967 with an installed capacity of 240 MW supplied by 24 turbine-generators, its tidal basin has an area of 22 km^2 with an average tidal range of 8 m and a peak of 13 m at spring tides. La Rance has been feeding electricity into the French grid for nearly half a century. Its historical importance is underlined by the fact that, until quite recently, the next largest scheme anywhere in the world was the 20 MW Annapolis scheme in Nova Scotia, Canada, installed in 1984 in a sub-basin of the famous Bay of Fundy which has the largest tidal range in the world. Then, in 2011, a 254 MW tidal barrage at Sihwa Lake in South Korea finally

Figure 1.16 A historic tidal mill in Portugal (Wikipedia).

Figure 1.17 The tidal barrage at La Rance in Brittany (Wikipedia).

overtook, by a very small margin, the pioneering French plant's generating capacity.

Large tidal barrages need good tidal ranges and extensive basins or lagoons for water storage, presenting unique civil engineering challenges. However, their recent history has been more influenced by financial and environmental concerns than purely technical ones. A barrage scheme requires a huge capital investment which may be hard to justify; and the environmental effects of disturbing the natural flow of tides in and out of large tidal basins may be severe and hard to predict with any certainty. Few projects better illustrate such difficulties than the long-heralded scheme for the Severn Estuary on the west coast of England, which has the second largest tidal range in the world. It has been discussed for over a century, actively researched and planned since the first 'oil shock' in 1973, and more recently put on a UK government shortlist for action. The Severn Barrage would be huge, with gigawatts (thousands of megawatts) of installed capacity able to meet perhaps 5% of the UK's electricity demand. However, in 2011 it was again shelved by the UK government, criticised by economists for its capital cost and by environmentalists for its feared effects on the local ecology. Similar difficulties have, over many years, hindered further development of tidal barrage schemes in Canada's Bay of Fundy [6].

As already noted, tidal mills and barrages use the potential energy of water raised by tides, whereas tidal stream technology taps the kinetic energy of flowing water. A sort of 'half-way house' between the two approaches is provided by *run-of-river* schemes that harness the energy of rivers without restricting their flow. Small installations often tap a proportion of river water upstream, feed it through a pipe to a downstream turbine, and then return it to the river. Large ones normally include dams to provide a good head of water for the turbines, with spillway gates to pass any excess flow. They are often seen as more environmentally benign than conventional hydroelectric power plants that store massive quantities of water behind dams and flood large tracts of country upstream. However, electricity generation, which varies in sympathy with the seasonal flows of rivers that power it, tends to be far more variable. Even so there are currently more than ten run-of-river schemes in the world with installed capacities greater than 10 MW, including four in British Columbia, Canada. The largest scheme in the world, completed in 1979 with a huge installed capacity of 2620 MW, includes the Chief Joseph Dam on the Columbia River in Washington, USA (see Figure 1.18). Note that although the dam provides a head of water for the turbines, the upstream reservoir is relatively small and the river is not impeded. Spillway gates cope with any excess flow.

In many respects large run-of-river schemes, in spite of their name, are technically more akin to tidal barrages than tidal stream systems. Most

Figure 1.18 The Chief Joseph Dam on the Columbia River in Washington, USA (Wikipedia).

use dams to create a head of water and, in this sense, operate on potential rather than kinetic energy. So what are the historical antecedents of true tidal stream technology, which taps the ongoing flow of water directly? The answer is 'not many'. There have been occasional attempts to power water wheels by partial immersion in a vigorous river or tidal flow – an early example being the wheels installed under the arches of London Bridge in 1580 to pump fresh water around the city – and various schemes and patents have since aimed to produce mechanical energy by ingenious methods [6]. However, very little was heard of tidal stream technology through most of the twentieth century, probably because environmental concerns were considered less pressing than they are today. Even the first 'oil shock' of 1973, which produced a flurry of research activity in wave energy, had little practical impact.

So what makes tidal stream systems so attractive today, and why are engineers busy considering the installation of thousands of turbines below the surface of the sea in fast-moving water? There are several key factors:

- *Low environmental impact.* Large arrays of tidal stream machines are confidently expected to have far less environmental impact than tidal barrages. Most will be invisible, operating on or near the sea bed.

- *Scale and modularity.* The power rating of individual machines is likely to be in the range one to a few megawatts, with physical dimensions that, although impressive, will not make them unwieldy giants. Large tidal stream 'farms' can be built up as arrays of interconnected machines on the modular principle, giving great flexibility over the growth of installed capacity. In many ways such farms will be analogous to wind farms, with their wide variations in size and number of individual turbines.

- *Technical.* Tidal stream technology – especially of the turbines that are expected to power most projects – has close connections with conventional hydroelectricity and modern wind energy.

- *Finance and economics.* A certain scale of power production is required for competitive generation into electricity grids, and machines rated at around 1 MW are widely regarded as the benchmark. Building up total capacity with such machines avoids the huge capital investment and risky economics of large tidal barrages.

All these factors are stimulating tidal stream technology (Figure 1.19). Its historical background may be modest, but it relies heavily on an understanding of the world's tides built up over many centuries. The tapping of tidal streams, a recent addition to the world's renewable energy portfolio, is now forging its own history.

1.4 Power, energy and performance

We have already referred to electrical power and energy a number of times. For example, when discussing wave resources in Section 1.2.1 we noted that the world total is of the order of 2–3 TW, and Figure 1.7 showed average values of wave power off the coasts of Western Europe expressed in kilowatts (kW) per metre of wave front. We have also referred to the installed capacity of tidal barrages, and the rated power of turbines, expressed in megawatts (MW). The word *energy* is extensively used in such terms as *wave energy* and *renewable energy*. In the public mind, power and energy may be more or less synonymous, but in science and technology they are distinct – and if we are to appreciate the contribution that wave and tidal devices can make to electricity production, it is essential to understand the distinction. We also need to be familiar with the various units used to measure them.

A key issue is the variability of electrical generation by wave and tidal devices, which impacts greatly on overall performance. Many people

Figure 1.19 A tidal stream turbine rated at 1 MW (Voith Hydro GmbH).

are confused about how much electricity the devices actually produce, because they are aware that seas are sometimes calm and tidal streams ebb and flow. Professional engineers and scientists are quite used to dealing with variability, understanding that electrical output can only be sensibly discussed in terms of statistical averages over significant time scales, but the general public needs assurance that today's technology is what it claims to be. Any uncertainty is played upon by climate-change sceptics, many with vested interests in opposing renewable energy.

We will tackle such issues by relating the output of wave and tidal stream devices to the electricity consumption of households. However, before we start it is important to realise that large national electricity grids are not concerned with the variable output of individual devices, or even of arrays unless they are very large. As the technologies develop, what matters is their *total* generation, spread over wide sea areas experiencing different conditions of wave and tide. Although total generation is certainly variable, it is not 'on–off' like that of an individual device, and any intermittency is a minor concern to the grid system as a whole.

To illustrate the relationships between power, energy and performance, we will start with a 1 MW tidal stream turbine. How does its power rating relate to the amount of electricity it actually feeds into the grid? A key point is also an obvious one: the turbine only generates its full rated power of 1 MW when the tidal stream reaches, or exceeds, a certain speed. Much of the time it produces considerably less, as the tide ebbs and flows. Given a good location the turbine's *average* power output, measured day and night over a complete year, might typically be 35% of its rated maximum, in other words 0.35 MW. The precise percentage, referred to as the *capacity factor* or *load factor*, depends on the technical efficiency of the turbine and the tidal stream resource where it is located. Similar considerations apply to wave energy converters.

Developers and electricity companies rarely mention capacity factors, probably regarding them as too technical for the general public. Instead they try to relate technical performance to personal experience by comparing the amount of electricity produced against the consumption of households. For example, it might be claimed that a 1 MW tidal turbine meets the needs of 600 households. No doubt this is a reasonable way of explaining things to consumers, but it is obvious that the turbine cannot supply the households on an hour-by-hour, day-by-day basis because its output varies in sympathy with the tides. Supply and demand are often out of step and this can lead to confusion. It would be more accurate to say 'over an average year the 1 MW turbine generates electricity equivalent to the annual consumption of 600 households'. But this is rather a mouthful and its subtleties would probably be lost on most people.

In any case there are several reasons to treat such estimates with caution:

- The annual electricity production of a turbine (and therefore the number of 'households equivalent') is site-dependent. Within a turbine array, not all machines have exactly the same yield.

- Small but significant power losses occur during transmission, especially when turbines are placed far from consumers.

- Consumption of electricity by households tends to increase year-by-year as living standards rise.

- Consumption patterns within a country often vary considerably from region to region, and between city and rural communities (so which 'households' are being used in the calculation?).

- It is also worth noting that average household electricity consumption varies greatly from country to country. For example the average USA figure is roughly twice that of Western Europe.

We see that estimates of 'households equivalent' are necessarily approximate. This is not to say they are wrong, simply that they rely on certain assumptions and are subject to statistical variation. They give an easily understood indication of performance, but must be interpreted sensibly.

As we have said, it is important to distinguish between power and energy. In the discussion that follows we must remember that power is a *rate* of energy production or consumption and, therefore, has dimensions of (energy/time). Conversely, energy has dimensions of (power × time). To take a familiar example, if an electric heater is rated at 1000 W or 1 kW, this is the amount of *power* it consumes when switched on. If left on for 1 hour, the *energy* used is 1 kWh (generally referred to as 1 'unit' of electricity). A 2 kW heater switched on for half an hour also uses 1 kWh of energy – power is being consumed at twice the rate but for only half the time. As householders it is energy we pay for, expressed in kWh or units of electricity. When assessing the contribution made by wave and tide machines to electricity generation, the key quantity is not the power they generate from one minute to the next, but rather their *annual energy production*.

We may express the annual energy production of a machine in terms of its rated power and capacity factor. For a turbine rated at P_r MW operating at a capacity factor C_f the average power output, measured over a complete year, is given by:

$$P_{av} = (P_r \times C_f) \text{ MW} \tag{1.1}$$

Since there are 8760 hours in a year the turbine's annual energy production E_a is:

$$E_a = (8760 \times P_{av}) \text{ MWh}$$

$$= (8760 \times P_r \times C_f) \text{ MWh} \tag{1.2}$$

For example, a machine rated at 1 MW, operating at a capacity factor of 35%, is expected to generate about $8760 \times 1 \times 0.35 = 3066$ MWh per year.

Alternatively, the capacity factor may be estimated if we know the turbine's rated power and the annual amount of energy it produces:

$$C_f = E_a/(8760 \times P_r) \tag{1.3}$$

As we move up the power scale from individual devices to large arrays, and then on to national electricity production, the numbers increase dramatically and it is often more convenient to work with gigawatts or even terawatts. The various units for measuring power are related as follows:

1000 W = 1 kW (kilowatt)	1000 kW = 1 MW (megawatt)
1000 MW = 1 GW (gigawatt)	1000 GW = 1 TW (terawatt)

There is a corresponding set of units for energy, measured in kilowatt-hours (kWh), megawatt-hours (MWh), gigawatt-hours (GWh) and terawatt-hours (TWh).

A good example of the way information is presented to the general public is given by the Isle of Islay project, an array of ten 1 MW tidal turbines in the narrow channel separating the islands of Islay and Jura off the west coast of Scotland (see Figure 1.11). Technical details of the project will be given in Section 5.2.1.

When first announced in 2011, the BBC reported that Islay's 10 MW turbine array would produce 'enough electricity to power more than 5000 homes'. The Daily Telegraph newspaper agreed, adding that this was 'more than double the number of homes on Islay'. The basis of the calculation was not given, but it implied that the electricity needs of Islay would easily be met by the turbine array – ignoring the fact that household consumption normally accounts for only about a third of the total (for a start, there are seven whisky distilleries on the island and all are substantial users of electricity!). The Islay Energy Trust, a local charity based on the island, got nearer the truth by estimating the turbines' output as about 30 GWh of electricity per year, adding that it was 'about the same amount as is consumed annually on Islay and Jura'. Since Jura's population is only about 200, this was tantamount to equating the annual output to Islay's total electricity demand; yet it omitted to mention that turbine output would fluctuate in sympathy with the tides, and that a grid connection would be needed to ensure continuity of supply. All of which illustrates the difficulty of deciphering information presented to the public.

However if we accept the annual production figure of 30 GWh (30 000 MWh) quoted by the Islay Energy Trust, Equation 1.3 gives the capacity factor of the 10 MW array as $30\,000/8760 \times 10 = 0.34$, or 34%. And if the array meets the annual electricity needs of about 5000 homes, it implies an average consumption of 6 MWh per home, which is probably about right for a Scottish island community.

We will now summarise the above discussion and broaden it to give an overall picture. Figure 1.20 shows some typical power and energy figures for electrical consumption and generation in Western Europe. Power is expressed either as a rated (peak) value, or as an average measured over a complete year. Energy is shown as an annual total. The various items are:

- *Household.* The average power consumption of West European households, measured night and day over a complete year, is about 0.6 kW. Since there are 8760 hours in a year, this corresponds to an annual energy consumption of about $0.6 \times 8760 = 5256$ kWh = 5.3 MWh. (Peak power consumption depends on how many appliances are switched on simultaneously and is not normally of great interest – provided the household's fuses are not tripped!)

- *1 MW tidal turbine.* The peak (rated) power is 1 MW and assuming a 35% capacity factor the average power is 0.35 MW, producing annual energy of $8760 \times 1 \times 0.35 = 3066$ MWh = 3.066 GWh. This is equivalent to the annual electricity requirements of about $3066/5.3 = 578$ households.

- *10 MW turbine array.* With a peak power of 10 MW and 35% capacity factor, the average power is 3.5 MW giving an annual 30.66 GWh, equivalent to about 5780 households.

- *1 GW conventional power plant.* Large modern power plants (fossil fuel or nuclear) are typically rated between 1 and 2 GW. This is their peak power and also, in principle, their average power assuming continuous operation at maximum output. However, in practice, their capacity factors are less than 100% and we will use 90% as a typical figure. So in a full year the output of a 1 GW plant is about 8760×0.9 GWh = 7.9 TWh, equivalent to the needs of about $7\,900\,000/5.3 = 1\,500\,000$ West European households. This puts into perspective the challenge of substituting wave or tidal stream power for a conventional power plant: about 2600 machines, each rated at 1 MW, would be needed to produce annual electricity equivalent to a 1 GW conventional power plant.

Note that we have given the values in Figure 1.20 to two significant figures. Generally speaking the accuracy of calculations, and of statistical data relating to household electricity usage, does not justify more (unfortunately values are sometimes given to four or five significant figures, implying unwarranted accuracy).

Figure 1.20 assumes electricity generation by tidal turbines, but exactly the same arguments apply to wave energy devices. Of course waves and tides do not obey our orders, and there is often a mismatch between electrical supply and demand. This is why it is so important to distinguish between a device's peak power and its average power over a complete year. Variability, and to some extent intermittency and unpredictability, are features of all renewable

CONSUMPTION		Power		Annual Energy (E_a)	Capacity Factor (C_f)	Households Equivalent
		rated (P_r)	av^{ge} (P_{av})			
household			0.6 kW	5300 kWh = 5.3 MWh		1
GENERATION						
1 MW tidal turbine		1 MW	0.35 MW	3100 MWh = 3.1 GWh	35%	580
10 MW turbine array		10 MW	3.5 MW	31 GWh	35%	5800
power plant		1 GW	0.9 GW	7900 GWh = 7.9 TWh	90%	1 500 000

Figure 1.20 Electricity consumption and generation.

energy technologies that harness natural energy flows in the environment. On the whole, tidal stream energy is a lot more predictable than wave, wind or solar. But whatever the modality, when electricity demand exceeds supply, the shortfall must be made up by other forms of generation, such as fossil-fuel, nuclear, conventional hydro or biomass. One of the major challenges facing wave and tidal stream engineering will be to integrate substantial amounts of variable generation successfully into grid networks supplying a wide range of industrial and domestic consumers who know very little about waves or tides!

The potential of wave and tidal stream machines for reducing carbon dioxide emissions is often mentioned in the press and on various web sites but this turns out to be an even trickier issue than estimating 'household equivalents'. Certainly, the electricity generated offsets or replaces electricity produced by other means. But which means? In a country such as Norway with its abundant supplies of conventional hydropower, will new wave or tidal stream capacity replace hydroelectric generation, with its very low attendant carbon emissions? In India and China, currently burning a huge amount of coal in conventional power plants, will more renewable generation mean less coal burning? Many countries produce electricity from a range of fossil fuels, nuclear power and renewables. The environmental benefits of renewable generation must clearly depend on the energy strategy and 'energy mix' of the country concerned. No wonder the claims made for carbon dioxide reductions by renewable technologies differ widely. It is a lot simpler to stay with comparisons based on electricity generation!

1.5 Into the future

Recent years have seen many types of wave and tidal stream device progress from scale models to full-scale prototypes deployed at sea, and the industry is already building up valuable experience in designing for ease of maintenance, reliability and survivability in extreme marine environments. The coming years will surely see increasing numbers of multi-device arrays installed, commissioned and grid-connected. By the 2020s there is a good chance that arrays rated at 100 MW and above will be harnessing wave and tidal power on a large scale.

However, the detailed timing of all this is hard to predict because it depends on a number of factors which must come together to produce coherent national strategies, actively and consistently supported by governments. Key ingredients are:

- Capital support for development of multi-device arrays and systems, together with revenue support for the electricity generated.

- Timely provision of grid access at reasonable cost in coastal areas with high marine energy resources.
- Efficient planning and consent procedures, taking into account the environmental impacts of wave and tidal stream installations.

Not surprisingly, advocates of marine energy take an optimistic view of the time and effort needed to get all these conditions in place; sceptics point to stop-go policies pursued by many governments which have impeded the development of other renewable energy industries over the years. Perhaps we should take a middle-of-the-road view: that although the way ahead will probably be bumpy, marine energy has already achieved an unstoppable momentum.

What scale of installation will signify that the industry has really taken off and is starting to contribute significantly to electricity generation? A nation's progress is normally expressed in terms of its cumulative installed capacity (CIC) in megawatts (MW), equal to the total rated power of all its wave and tidal stream machines. Taking the UK as an example, the 2010 figure was just a few megawatts, with some experts predicting 50 MW by 2015 or soon after. However, this is a mere drop in the ocean compared with the nation's average electricity generation of around 50 GW. To make a significant contribution would require at least (say) 1 GW of CIC, which could be met by one thousand 1 MW machines or their equivalent.

How long will it take the UK to reach gigawatt scale? Assuming the key ingredients mentioned above are well and truly in place and that the appetite of developers and installers remains healthy, the industry's trade association Renewable UK places the date around 2020 [12, 13]. By this time global installed capacity might be a few gigawatts, but after that progress becomes ever harder to predict. Noting that the UK's practically extractable resource is about 36 GW, Renewable UK sounds a warning note:

> To get to the stage at which the UK is deploying marine energy arrays at the scale of 100 MW the industry must have de-risked the technology, mastered installation and operation and maintenance techniques, obtained greater understanding of environmental impacts, developed a supply chain and secured significant sums of private investment.

Another respected UK organisation, the Carbon Trust, has gazed much further into the crystal ball and made separate predictions for wave and tidal stream capacity by the year 2050, based on various development scenarios [14]. Its most optimistic figures for the UK are up to 20 GW for wave energy and 8 GW for tidal stream; and for global capacity, up to 189 and 50 GW, respectively. All these figures are substantially reduced in less optimistic scenarios and the Carbon Trust even acknowledges the possibility that marine energy may, in the next 30 years, fail to achieve real take-off if governments are distracted by other energy options.

Given all the uncertainty, it is worth reflecting on progress in the global wind and solar industries. Since reaching the 1 GW landmark (in 1993 and 2000, respectively) both industries have doubled in CIC about every three years, suggesting that today's renewable energy industries, increasingly supported by governments around the world, can achieve impressive growth. This is an observation rather than a prediction for marine energy, which has its own special features, but there are certainly grounds for optimism. Of course it is unclear how the global manufacturing and installation of wave and tidal stream machines will be shared between today's major players – principally the UK and other countries in Western Europe, plus the USA and Canada.

Figure 1.21 suggests two possible scenarios for growth of the UK's CIC in wave and tidal stream generation. The two technologies may contribute about equally, although some experts predict that tidal will initially outpace wave due to the latter's special engineering challenges. Curve A is broadly in line with Renewable UK's predictions – reaching 1 GW by about 2020 and continuing rapidly towards 10 GW and beyond. Curve B takes a more sober view, assuming a doubling in CIC about every three years and reaching 1 GW in the early 2030s – the sort of progress actually achieved by the global wind and solar energy industries once they reached the point of take-off. Note that both 'curves' are actually straight lines on a semi-logarithmic plot, implying exponential growth. This may be a reasonable assumption for a vibrant new industry in its early stages, but it must sooner or later tail off as market penetration increases.

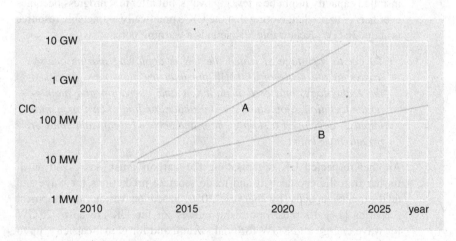

Figure 1.21 Potential growth of cumulative installed capacity (CIC) in the UK.

Whether one prefers scenario A or B – or something in between – it seems likely that marine energy companies will be employing tens of thousands of people by the 2030s. Of course this assumes that designers and engineers will continue to develop and refine their devices, proving that large arrays are viable technically, economically and environmentally. In the case of wave devices, issues of reliability and survivability will be especially important. It is certainly a big agenda, but the signs are good. In Chapters 4 and 5 we will meet many examples of devices at the forefront of development, which are leading the industry towards commercialisation and the gigawatt era.

References

1. E.F. Schumacher. *Small is Beautiful*, 1st edn, Vintage Books: London (1993).
2. A. Gore. *An Inconventent Truth*, Bloomsbury Publishing: London (2006).
3. H. Scheer. *A Solar Manifesto*, James & James: London (2005).
4. M.E. McCormick. *Ocean Wave Energy Conversion*, Dover Publications: New York (2007).
5. J. Cruz, ed. *Ocean Wave Energy: Current Status and Future Perspectives*, Springer: Berlin (2008).
6. R.H. Clark. *Elements of Tidal-Electric Engineering*, Wiley-Interscience: New York (2007).
7. J. Hardisty. *The Analysis of Tidal Stream Power*, Wiley-Blackwell: Chichester (2009).
8. S. Salter. Wave power. *Nature* **249**, 720–724 (1974).
9. S. Salter. *Looking Back*, Chapter 2 in ref. 5 above, Springer: Berlin (2008).
10. D. Ross. *Energy from the Waves*, Pergamon Press: Oxford (1979).
11. J. Falnes. *Ocean Waves and Oscillating Systems: Linear Interactions Including Wave Energy Extraction*, Cambridge University Press: Cambridge (2005).
12. RenewableUK (2011) Wave and Tidal Energy in the UK: State of the Industry Report, London.
13. RenewableUK (2012) Marine Energy in the UK, London.
14. Carbon Trust (2011) Marine Renewables Green Growth Paper, London.

2 Capturing marine energy

In the previous chapter we saw how designers and engineers have tried to capture the energy in waves and tides over several centuries – in the case of waves with a wide variety of machines, some stranger than others; in the case of tides, with tidal mills and barrages and, more recently, turbines that tap the kinetic energy of tidal streams. In this chapter we introduce the physics of ocean waves and tidal streams as a prelude to explaining the principles underlying some of today's most promising developments.

2.1 Ocean waves

2.1.1 Linear waves

The theory of linear waves, often referred to as *Airy waves*, was originally devised by the English mathematician and astronomer George Airy (1801–1892). A man of extremely wide interests in science and engineering, he established the prime meridian of longitude at Greenwich in London, was highly active in astronomical research, and made major contributions to structural mechanics and fluid dynamics – including a linear model to explain the form and propagation of surface waves on the world's oceans.

Wind-generated ocean waves in deep water are well described by Airy's theory, especially when they are built up gradually over a long distance (fetch) to form a regular, low-amplitude, swell [1, 2]. Figure 2.1 shows a 'snapshot' of a few linear waves at a particular instant in time. They are sinusoidal in form with amplitude A, height H between crest and trough,

Electricity from Wave and Tide: An Introduction to Marine Energy, First Edition. Paul A. Lynn.
© 2014 John Wiley & Sons, Ltd. Published 2014 by John Wiley & Sons, Ltd.

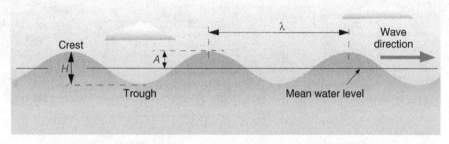

Figure 2.1 Linear ocean waves.

and wavelength λ. Although no mathematical model describes ocean waves exactly, the sinusoidal approximation works well when the wave height of a swell is much smaller than its wavelength. Unless otherwise stated, the discussion and results in this section apply to such waves, travelling over deep water and unaffected by the sea bed.

As waves travel across the ocean surface and pass a particular point in space, it is possible to measure the time interval between successive crests, known as the *period T*. The inverse of the period is the *frequency f*, and denoting the wave velocity by c we may write:

$$T = 1/f = \lambda/c \tag{2.1}$$

Airy's theory predicts that the period, wavelength and velocity of linear waves in deep water are interrelated by two further equations:

$$\lambda = gT^2/2\pi \tag{2.2}$$

and

$$c = gT/2\pi \tag{2.3}$$

where g is the gravitational constant ($9.81 \, \mathrm{m\,s^{-2}}$). The following table gives a few representative values:

Period T (s)	Wavelength λ (m)	Velocity c (m s^{-1})
2	6.2	3.1
5	39	7.8
10	156	15.6

It is perhaps surprising that the three variables are so tightly related. For example, linear waves with a period of 10 s ('10-second waves') *must* have

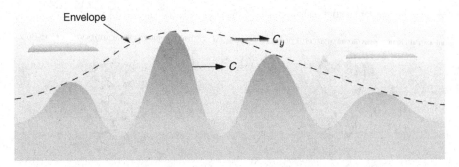

Figure 2.2 Part of a wave packet, showing phase and group velocities.

a long wavelength of 156 m and travel at a high speed of 15.6 m s^{-1}, independent of their height.

The above waves form part of a continuous, ongoing, sequence. However, it is also possible for a small group of waves to be set up and travel as an isolated *wave packet*. A good example is the bow wave fanning out from a moving ship, which typically forms a wave packet with a few wave crests [1]. Close observation reveals that individual waves are born at the back of the group, move steadily to the front, and then die, leaving the *envelope* unchanged, see Figure 2.2. This shows that individual waves move faster than the wave packet as a whole; and if the ship is fairly close to shore, it is the velocity of the wave packet, not individual waves, that determines how long the energy will take to reach land and dissipate itself against the shoreline.

It turns out that the deep-water velocity of individual waves – their *phase velocity* (c) – is exactly twice the velocity of the group as a whole, known as the *group velocity* (c_g). Thus:

$$c_g = c/2 = gT/4\pi \tag{2.4}$$

This result is important for wave energy conversion because energy transport across an ocean surface takes place at the group, not the phase, velocity. And since the group velocity is proportional to the period it follows that low-frequency waves move more rapidly away from a storm centre, reaching any distant wave energy converters (WECs) before their high-frequency cousins.

Figures 2.1 and 2.2 illustrate surface profiles of waves. But what happens beneath the surface and, in particular, how do the myriad water particles that make up a wave actually move? First, we must realise that it is energy, not water, that travels across the surface of an ocean. There is no net

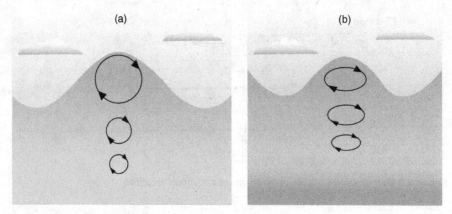

Figure 2.3 Water particle movements.

translation of water particles; instead, they simply move in small vertical circles as each wave passes, the circle radius decreasing exponentially with depth, see Figure 2.3a. The decrease is so rapid that virtually all energy transport takes place within half a wavelength of the surface. The energy is partly *potential*, due to the height of the waves; and partly *kinetic*, due to movement of the water particles.

The idea of energy transport occurring without an equivalent mass transfer of water may seem strange. But imagine a long rope laid out on the ground: pick up one end, give it a sudden jerk, and a ripple travels all the way to the other end. It is energy, not rope, that is being transported horizontally.

As waves move towards shallow water, the circular paths of water particles change gradually into ellipses [1]. At intermediate depths the major and minor axes of the ellipses decrease steadily towards the sea bed, see Figure 2.3b. In very shallow water the ellipses have constant major axes but decreasing minor ones. A point is reached where the waves' phase velocity, which reduces with water depth, equals the maximum horizontal velocity of the water particles, causing the waves to *break*. Their tops roll over and they end up dissipating energy in friction and turbulence.

Another important phenomenon, known as *shoaling*, occurs when unbroken waves enter shallow water. They increase in height and reduce in wavelength (although the frequency remains unchanged). The basic reason for this is that the group velocity, which is also the energy transport velocity, decreases in shallow water. However, the rate of transfer of energy (the energy flux) remains essentially constant as long as the waves do not break, so the reduction in transport velocity must be compensated by an increase in wave height.

Breaking waves obviously cannot be modelled by a linear theory and sinusoidal wave profiles. In fact the linear approximation becomes inaccurate as waves start moving into shallow water, before any breaking occurs. Typically their profile changes to a narrower crest and broader trough than shown by Figure 2.1. Various theories have been developed to describe the properties of nonlinear waves although, not surprisingly, they are relatively complicated [1].

We have already noted that wave energy is transported at the group velocity. But how much energy? The usual way of expressing this is by a parameter J equal to the number of kilowatts per metre of wave front, and linear theory gives the following result for deep water [2]:

$$J = \rho g^2 T H^2 / 32\pi = 0.986\, T\, H^2 \text{ kW m}^{-1} \tag{2.5}$$

Here ρ is the specific density of sea water (1.03), g is the gravitational constant (9.81 m s^{-2}) and, as before, T and H are the period and height of the waves, respectively. It is important to note that the energy transported by waves is proportional to the square of their height.

Since 0.986 is very close to unity, for all practical purposes we may write Equation 2.5 as:

$$J = T H^2 \text{ kW m}^{-1} \tag{2.6}$$

This easily remembered result shows that, for example waves with a period of 10 s and height 2 m transport 40 kW m^{-1} across an ocean surface. It is interesting to refer back to Figure 1.7 showing typical wave power values off the coasts of Western Europe. These fall in the range 40–70 kW m^{-1} although they are, of course, long-term averages produced by waves of widely differing periods and heights throughout the year.

We have now introduced some key properties of ocean waves. Airy's theory is certainly helpful for understanding the basic mechanisms of wave formation and transport, and its results are realistic for swells in deep water built up over long stretches of ocean. But we must always remember that linear theory assumes idealised sinusoidal waves, whereas practical wave energy conversion must allow for 'awkward' waves in real-life seas.

2.1.2 Random waves

The science of ocean waves would be a lot simpler if all waves were linear – sinusoidal in shape, constant in height and wavelength, with easily calculated values of phase and group velocities. So too would be the design of WECs which, in most cases, work best in regular, predictable swells arriving over long stretches of ocean.

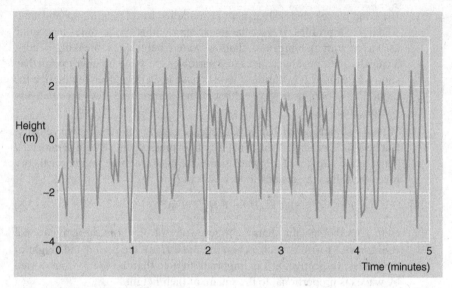

Figure 2.4 Wave heights in a rough sea.

But anyone who has experienced a storm off the northwest coast of Scotland or in the notorious Bay of Biscay knows that real-life ocean waves are far more complicated. Airy's linear theory, undoubtedly useful for understanding gently oscillating ocean swells, hardly addresses the reality faced by WECs: fickle and variable sea states, wave profiles often far from sinusoidal and occasional waves high enough to test the most robust devices.

Figure 2.4 shows a typical 5-minute record of wave heights in a rough sea off the west coast of Scotland. The wind is generating choppy local waves superimposed on a long-distance ocean swell. Individual heights vary between about 3 and 6 m from trough to crest, with occasional waves well outside this range. Looking carefully, we see that there is a fairly constant underlying frequency of about 6 waves per minute, or one every 10 s, corresponding to the swell; but the locally generated waves give the pattern a random appearance and it bears little resemblance to the neat sinusoidal profile of Figure 2.1.

Clearly, the designers of WECs need to describe and understand such wave patterns, preferably without getting involved in obscure mathematics. As we saw in the previous section, the regular waves of a gentle swell conform closely to the sinusoidal model with a well-defined height and

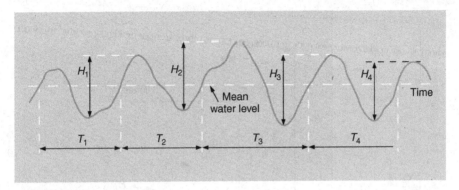

Figure 2.5 Measuring random waves.

period, but random waves are more awkward and can only be described in statistical terms.

The most important statistical measure of random waves is the *significant wave height (SWH)*, traditionally defined as the average height of the highest third of waves. Figure 2.5 shows individual heights H_1 to H_4 of a group of four waves. In a rough sea many such measurements are needed to get a reliable SWH estimate – separating out the highest third and calculating their average height. For example, if 150 waves are measured the average height of the 'top 50' must be calculated.

The above definition of SWH was formulated during World War II as a convenient, readily calculated, measure of a sea's roughness. It was designed to agree with visual estimates by 'trained observers' and became widely accepted by meteorologists and mariners. However, it is rather awkward from the physicist's point of view because it does not relate easily to standard statistics or to modern ways of recording wave heights with automatic equipment. An alternative is to use a *Rayleigh distribution* to model the probabilities associated with the full range of wave heights encountered in random seas.

English physicist Lord Rayleigh (1842–1919), a Cambridge professor and Nobel Prize winner with a polymath's interests in the physical sciences, originally developed his probability distribution to describe the scattering of radio waves. It has since found widespread application in the physical sciences; for example it is often used to describe the variability of winds that drive wind turbines. It takes the form:

$$p(x) = (x/\alpha^2) \exp(-x^2/2\alpha^2) \tag{2.7}$$

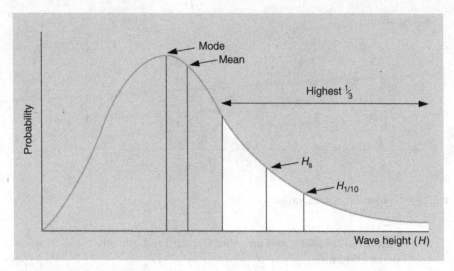

Figure 2.6 A Rayleigh distribution of wave heights.

where p is the probability associated with a value x and α is a constant. Figure 2.6 shows it applied to ocean waves and we should note the following points [3]:

- The graph shows the relative probabilities, or occurrences, of different wave heights. There are relatively few small waves (left side of graph), and a small number of very large waves (far right of graph). Most waves fall in the mid range of heights.

- The most probable wave height is the *mode*, denoted by α in Equation 2.7; the average wave height is the *mean;* and the height exceeded by 10% of waves is denoted by $H_{1/10}$.

- The highest third of waves is represented by the white shaded area, equal to one third of the total area under the curve. The average height of waves in this shaded group is the SWH H_s.

Assuming a Rayleigh distribution, the SWH is easily related to standard statistical measures, such as the mean, mode and standard deviation (σ) because they are all proportional. For example, H_s is very close to $4\,\sigma$ and, as we shall see in the next section, the SWH is nowadays usually defined in this way, giving results very close to those calculated by the traditional method.

As a practical example, suppose a marine weather forecast predicts a stormy sea with a SWH of 10 m. Assuming a Rayleigh distribution of wave heights, we may expect the following approximate values [3]:

- A mean (average) wave height of 6.4 m
- A mode (most likely) wave height of 5.1 m
- 10% of waves to exceed 12.7 m
- 1% of waves to exceed 16.7 m.

The most important conclusion is that a WEC exposed to stormy seas must expect to meet waves far higher than the SWH. One wave in a hundred may be 70% higher, and twice as high is a distinct possibility (producing what is popularly called a *rogue wave*). Such statistics are well understood and respected by experienced mariners and they give plenty of food for thought to the designers of wave energy machines.

Talk of high waves leads naturally to a brief discussion of *breaking waves*. We have all seen waves break as they approach a gently-sloping beach, and noticed the 'white horses' that appear in a rough sea offshore. The reason a wave breaks is that its base can no longer support its top, leading to collapse. Whenever the *steepness* of a wave, defined as its height H divided by its wavelength λ, becomes too great a collapse is inevitable. Breaking occurs in three common situations:

- In shallow water where the depth d is much smaller than the wavelength, if the wave height exceeds about $0.8d$.
- In deep water, if the steepness exceeds about 0.17.
- In wind-blown water, if the wind is strong enough to blow the top off the base.

The breaking of waves is a highly complicated nonlinear process, arguably one of the least understood phenomena in ocean science.

Another important property of random waves is their *period* which, like the height, is a statistical variable. Referring back to Figure 2.5 we see four consecutive values labelled T_1 to T_4 measured as the interval between consecutive upward (it could also be downward) crossings of the mean water level. The *significant wave period* (*SWP*) is defined as the average period of the highest third of waves, mirroring the traditional definition of the SWH. So once again, if 150 waves are measured, the periods of the highest 50 are separated out and averaged. Alternative ways of estimating the period are often used nowadays and will be discussed in the following section.

So far we have not discussed the frequency of random waves. In the case of linear waves the frequency is simply the inverse of their period – see Equation 2.1. With random waves the concept of frequency, like that of height and period, is less easy to visualise but, as we shall see in the following section, it forms the basis of a powerful set of techniques for describing random seas and their practical exploitation.

2.1.3   Wave spectra

The height, period and frequency of surface waves are related to wind velocity and direction. A light local breeze generates waves of small height and period that generally travel in the same direction as the breeze; but waves generated in storm conditions tend to be much higher, with longer periods, and travel in many directions. Real-life seas encountered by WECs are often very complex.

So far our story has progressed from linear waves with their sinusoidal profiles and well-defined properties to random waves described by simple statistical measures. We have concentrated on the two most easily observed properties of ocean waves – height and period. But it is now time to tackle a third property, *frequency*, which proves especially valuable for describing complex random seas. We are about to move into the *frequency domain*.

The frequency of sinusoidal waves is easy to define and understand; it is simply the reciprocal of the period. For example, a continuous sequence of linear waves with a period of 10 s has a frequency of 0.1 waves per second. A random sea, however, may be thought of as the superposition of a large number of waves with different heights and frequencies, travelling in many directions. This is the essential insight provided by *Fourier analysis*, brainchild of the French mathematician and physicist Jean Baptiste, Baron de Fourier (1768–1830), who showed that a complex waveform can be considered as a sum of sinusoids of appropriate magnitude, frequency and phase, referred to as its *spectrum*. Fourier analysis has extensive applications in many branches of engineering science including electronics, communications and mechanical vibrations, and the wave spectrum is nowadays accepted as a powerful tool for describing a random sea (Figure 2.7).

The wave spectrum tells us how much energy is carried by the various frequencies present in a random sea. It is often given the symbol $S(f,\theta)$ to emphasise that it is a function of both frequency and direction, referred to as the *directional spectral density*. But when considering a single direction it is written more simply as $S(f)$. Integrating this function over all frequencies gives the total amount of wave energy transported [2]:

$$\int S(f)\mathrm{d}f = H_\mathrm{s}^2/16 \qquad (2.8)$$

Here H_s is the SWH defined in the frequency domain which, in most cases, gives values that agree well with those discussed in the previous section.

Figure 2.7 Waves in a random sea (Paul A. Lynn).

For reasons discussed below, it is often written as H_{m0}. The equation shows how knowledge of the wave spectrum of a random sea allows us to estimate its SWH.

What do wave spectra actually look like? If the wind blows steadily for a long time over a large sea area, the waves reach an equilibrium with the wind, producing a *fully developed sea* with a well-defined spectrum. In this context a long time is typically several thousand wave periods, and a large area is several thousand wavelengths across. Such stable conditions are rare; more typically, strengthening winds build up a *developing sea*, which does not have the chance to reach equilibrium before winds subside – or change direction. In its initial stages a developing sea exhibits mainly low waves of short period, but these gradually turn into high waves of long period. This means that the spectrum of a random sea tends to be quite variable; but under normal conditions it is stable enough during the 30 minutes or so needed to measure it.

Figure 2.8 shows some typical shapes of wave spectra in random seas – illustrative rather than definitive and designed to show some broad effects:

- Figure 2.8a is for an ocean swell that has travelled a long distance from its region of generation, producing fairly regular waves with a period of about 10 s. If the waves were strictly linear (sinusoidal) all the spectral energy would be concentrated at a single frequency of

49

about 0.1 Hz, but in practice variability causes some 'spreading' of the spectrum.

- Figure 2.8b represents a sea with two main types of wave: a long-distance swell similar to that in Figure 2.8a, plus some substantial wind-waves generated more locally with shorter periods and higher frequencies.

- Figure 2.8c shows three stages of a developing sea. In stage 1 rising winds produce waves with relatively high frequencies, by stage 2 most waves have longer periods and lower frequencies, by stage 3 the sea has become well developed, with dominant low-frequency waves.

As already noted, defining the SWH in the frequency domain reflects the increasing use of computer techniques and Fourier analysis in marine science. To understand the relationship between H_s and $S(f)$ specified in Equation 2.8 we need to add some further comments on wave spectra.

The integral of $S(f)$ in Equation 2.8 is known as the *zeroth moment* of the wave spectrum and is generally given the symbol m_0; and the SWH H_s is often written as H_{m0} to emphasise that it is defined in the frequency domain. In most random seas, and especially those displaying a fairly narrow spectral distribution (such as Figure 2.8a), the various ways of defining and measuring SWH are found to agree closely so that to a good approximation:

$$H_{1/3} = H_s = H_{m0} \tag{2.9}$$

As far as H_{m0} is concerned, many investigations of random seas have shown that it is approximately equal to four times the square root of the zeroth moment m_0. Thus:

$$H_{m0} = 4\sqrt{m_0} \tag{2.10}$$

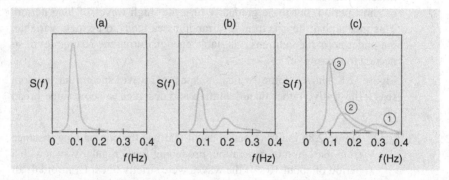

Figure 2.8 Typical wave spectra.

and therefore:

$$m_0 = H_{m0}{}^2/16 \qquad (2.11)$$

which ties in with Equation 2.8.

To summarise we have, by rather a tortuous route, progressed from the historical definition of SWH based upon subjective judgements by 'trained observers' ($H_{1/3}$) to a modern one suited to Fourier analysis and digital computation (H_{m0}). Equation 2.10 is now widely accepted as the definition of SWH in a random sea.

So far we have not considered the two-dimensional nature of wave spectra – their variation in direction as well as frequency. This is certainly an important issue for wave energy conversion, but it involves complicated mathematics and computation [4] beyond the scope of this book. We will settle here for the diagram in Figure 2.9, a *wave rose* measured by a directional buoy showing variations of wave energy and SWH as a function of direction, or azimuth (compass bearing). This particular wave rose is fairly typical of an ocean location in the north Atlantic, a few kilometres off a west-facing coast. The prevailing winds are westerly and most wave energy comes from roughly this direction; almost none comes from the east due to the scarcity of easterly winds and the presence of the coast. The coloured bars in each 15° sector indicate the relative amount of energy

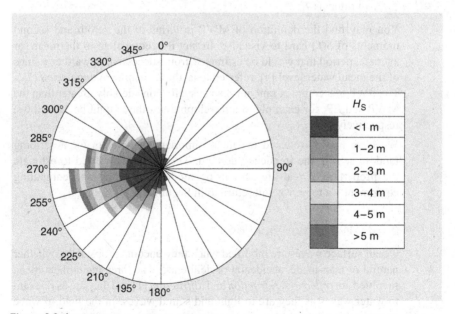

Figure 2.9 A wave rose.

received over a complete year at different values of H_s. Such information is critical for the design of WECs and their moorings, especially if the devices have pronounced directional characteristics.

The other key parameter in relation to wave spectra is the period of a random sea. Towards the end of the previous section we explained that a traditional measure, known as the *significant wave period*, is defined as the average period of the highest third of waves. But we are now working in the frequency domain so the question naturally arises: can we estimate the period of random waves from the spectrum $S(f)$? Indeed we can, and three measures are commonly used:

- The *peak wave period* (*PWP*) is the reciprocal of the frequency corresponding to the peak value of $S(f)$. Thus if the peak of $S(f)$ occurs at frequency f_p, the PWP equals $1/f_p$. The PWP is representative of the higher, more energetic, waves in a random sea.

- The *mean wave period* (*MWP*) is equal to $\sqrt{(m_0/m_2)}$, where m_0 is the zeroth moment of $S(f)$ as already discussed, and m_2 is the second moment of $S(f)$ defined as $\int f^2 \, S(f) \, df$. It is often written as T_{02} to indicate its frequency domain definition. The MWP is representative of the full range of periods in a random sea.

- The *energy period* (*EP*) is the period of a linear (sinusoidal) wave that would carry the same total energy as a random sea. It is often written as T_e.

You may find the definition of MWP in terms of the zeroth and second moments of $S(f)$ hard to visualise. In fact it is equivalent to the mean, or average, period that would be estimated from successive upward crossings of the mean water level [4], referred to as the *zero up-crossing period* (T_z). The PWP of a complex random sea is usually considerably greater than the MWP and EP; for example in a developing sea they might be 10 and 6 s, respectively.

We have now introduced the most important parameters for describing random seas in the frequency domain. They are widely used to estimate the properties of real-life seas using modern recording and computing equipment – a topic we will return to in Section 2.1.5.

2.1.4 Wave modification

Ocean surface waves are modified when they encounter obstacles, whether natural or man-made, accidental or deliberate. The three main phenomena involved are *reflection, refraction* and *diffraction* [1] and they are as relevant to water waves as they are to light and sound waves. In the field of wave energy conversion, they can affect the wave patterns encountered by wave

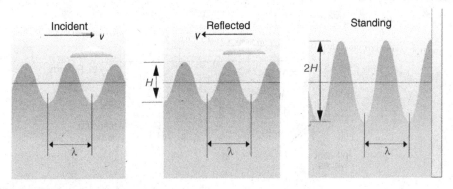

Figure 2.10 Formation of a standing wave.

energy machines, and the efficiency of the machines themselves. They are also very important in the design of wave tanks used to test scale models of new devices.

Reflection may be defined as a change in the direction of a wave front at a boundary between two different media, so that it returns to the medium from which it originated. Familiar examples are light falling on a mirror, and sound waves producing echoes. When ocean waves meet a vertical barrier, such as a sea wall, their energy is usually partly absorbed by the barrier and partly reflected. In the special case of linear waves meeting a smooth rigid barrier, all the energy is reflected back in the direction from which it came to produce a pure *standing wave*, illustrated in Figure 2.10. The incident and reflected waves have the same height and add together (superpose) to produce a standing wave twice as high. It has zero phase velocity since the phase velocities (*v*) of the incident and standing waves cancel one another. This makes it appear stationary, with individual crests and troughs waxing and waning. Its energy is twice that of the incident wave.

In Figure 2.10 the incident waves meet the reflecting barrier 'head-on'. The situation is more complicated, and harder to visualise, when they meet it obliquely. Figure 2.11 illustrates this with a plan view of waves meeting a long, smooth barrier at an angle θ. Crests are shown red, troughs blue. The angles of incidence and reflection are equal, and superposition of incident and reflected waves produces a 'checkerboard' pattern with crests (red dots) where the crests of incident and reflected waves coincide, and troughs (blue dots) where their troughs coincide. But unlike the standing wave in Figure 2.10, the whole pattern moves sideways along the barrier.

So far we have considered linear waves meeting smooth, rigid, vertical barriers and undergoing perfect reflection but things are not generally that

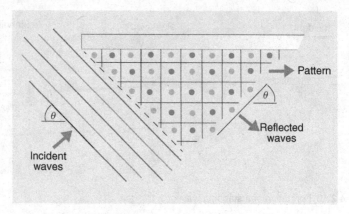

Figure 2.11 Oblique wave reflection.

simple. More commonly, random waves with less distinct patterns meet partially-reflecting barriers of irregular form. Yet the reflections, although more chaotic, may affect WECs because they can be produced by cliff faces, harbour walls and moored vessels.

Refraction causes a change in the direction of waves due to a change in their speed. A familiar example is the bending of light waves as they pass obliquely through a lens or prism. In the case of water waves refraction occurs as they move obliquely into water of a different depth, typically as they approach a beach. The effect begins when the depth d reaches about half the deep-water wavelength λ, causing a reduction in phase velocity. As the waves move further up the beach they bend gradually round to meet the shoreline. Figure 2.12 shows a single wave crest moving from deep water

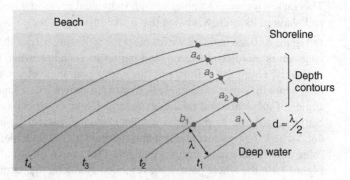

Figure 2.12 Refraction of a wave approaching a beach.

towards a sloping beach with straight, parallel depth contours. At instant t_1 point a_1 on the crest has just reached the depth $d = \lambda/2$. As it moves up the beach its phase velocity and wavelength decrease. At instants $t_2, t_3, t_4, \ldots$ it reaches point a_2, a_3, a_4, and so on. Point b_1 starts the same process at instant t_2. In this way increasing widths of the wave front enter shallow water and bend slowly round to meet the shoreline at a reduced angle.

As far as wave energy conversion is concerned, refraction tends to simplify the design and improve the performance of near-shore and shoreline devices by providing them with wave fronts that approach more or less parallel to the shoreline, meaning that they only have to extract energy coming from a narrow range of directions. There has also been considerable interest over the years in *focussing techniques* that use refraction to concentrate wave power, for example by making waves pass over the carefully shaped slopes of an artificial atoll [1].

Our final topic, *diffraction*, causes ocean waves to spread out on the far side of slits or openings, for example in a sea wall, especially when their size is comparable with one wavelength. It also causes waves to bend around obstacles. Figure 2.13 shows a plan view of these effects, and we see that waves can pass into calm water behind an obstacle even though they cannot 'see' the *shadow zone* they are about to enter. As incident waves clear a barrier such as the end of a sea wall the wave fronts tend to become circular, as though the barrier were itself a point source of wave energy. As far as wave energy is concerned, diffraction around obstacles – narrow spits of land, rock outcrops, breakwaters, harbour walls – may affect the energy reaching a wave device and, in principle, diffraction through a gap could be used to focus energy towards a point where a device is located [1].

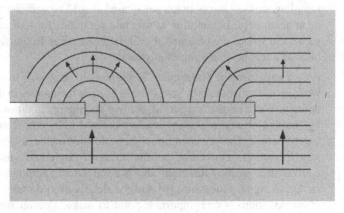

Figure 2.13 Wave diffraction at a gap, and at the end of a barrier.

2.1.5 Wave measurement

Modern wave measurement systems rely on sophisticated digital electronics for estimating and relaying parameters, such as wave height, period and spectrum back to the shore. In most cases the statistics of random seas are reasonably stable for long enough to allow accurate estimates to be made. The basic approach involves regular *sampling* of the surface water level over a specified time interval, typically about half an hour, followed by digital processing to calculate the required parameters.

If we refer back to the record of wave heights shown in Figure 2.4, an obvious question arises: how often should such a wave pattern be sampled? Too often, and we have a surfeit of data to process; too infrequently, and the digital representation of what are essentially analogue waveforms is not detailed enough. Intuitively, the higher the frequencies (and rates of change) present in a wave pattern, the more often we need to sample it – a point borne out by one of the most important ideas in digital communications – the *Nyquist sampling theorem.*

Harry Nyquist (1889–1976) was a Swedish-American engineer who made fundamental contributions to information and communications theory. His sampling theorem may be stated as follows:

> *A function containing no frequencies higher than f_{max} Hz may be completely represented by a series of samples spaced $1/2f_{max}$ seconds apart.*

The spectra previously illustrated in Figure 2.8 show that most ocean wave energy is contained in frequencies below about 0.5 Hz. To err on the safe side we might allow for components up to, say, 1 Hz. This implies that sampling surface waves twice per second should be sufficient to represent them accurately. If sampling at this rate continues for half an hour, 3600 numerical values are obtained for digital processing. In practice, sampling rates of a few hertz are commonly used.

As we explained in Section 2.1.3, the modern approach to computerised wave measurement makes heavy use of Fourier techniques in the frequency domain. Wave height and period, the two parameters traditionally assessed by 'trained observers', are nowadays derived indirectly from the energy spectrum $S(f)$. The key operation to be performed on the raw data is *Fourier transformation* – converting a time-series of numerical samples into an equivalent frequency function. This is a standard computational procedure in digital signal processing [6] and the details need not concern us here. However, there is one important point to make: Fourier transformation, which tends to be computationally slow for long sample sets, can be speeded up using a *Fast Fourier Transform* (*FFT*) algorithm. Such algorithms, which achieve their effect by eliminating redundant calculations, are invariably included in today's wave measurement software.

A variety of measurement buoys [4] are deployed in the world's seas and oceans, providing information for navigation, weather reporting, off-shore oil and gas exploration and wave energy research. Many modern devices measure wave direction as well as height and period, producing two-dimensional wave spectra; some combine wave and tidal current measurements in a single device; and various radio, cellular and satellite telemetry options are used to transmit data back to shore. We now introduce a buoy that illustrates the features and capabilities offered by today's wave measurement technology.

The Dutch company Datawell [5] started producing its *Waverider* buoys in 1961 and many thousands have since been deployed around the world. Initially the devices were restricted to measuring wave height and period, but *Directional Waveriders* were introduced in 1988 and have since undergone continuous development. The principal features of the Mark III version may be summarised as follows:

- The device takes the form of a spherical hull 0.9 m in diameter (see Figure 2.14).
- The wave motion sensor uses a stabilised platform, accelerometers and a magnetic compass.
- Wave heights are measured for wave periods between 1.6 and 30 seconds. Wave direction and water temperature are also measured.
- GPS (global positioning system) is used for buoy monitoring and tracking through a HF (high frequency) radio link.
- Options are available for satellite monitoring and tracking.

A recent version of the device, introduced in 2012, extends the capabilities with a solar panel for battery charging, and an acoustic current meter based on the Doppler effect for measuring surface current velocity.

Wave measurement buoys are used by marine energy organisations to provide the continuous, long-term, data required by wave energy developers as they deploy their machines offshore. In addition, a number of countries, including the USA, Canada, India and several in Europe, have installed networks of buoys around their coasts, some of which provide directional wave information.

A very different approach to wave measurement is provided by satellite altimeters [4]. At first sight it seems unlikely that a radar altimeter, travelling about 1000 km above the Earth's surface at a speed of about 6 km s^{-1}, could provide useful information about the waves below, but in fact the

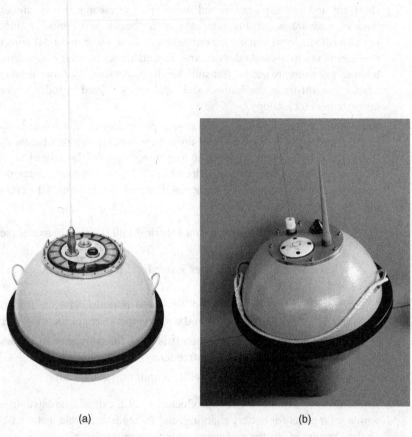

(a)   (b)

Figure 2.14 Directional *Waverider* buoys incorporating a solar panel (a) and an Argos satellite link (b) (Datawell BV).

radar echoes returned from ocean waves, expertly interpreted, can yield SWH measurements with an accuracy close to that of a buoy floating on the sea surface. Essentially, this is done by measuring the slope of the leading edge of the returning echoes. Satellite altimeters need careful calibration, normally achieved by comparing their data with that obtained from offshore buoys, preferably well away from coasts where wave patterns are reasonably stable, both temporally and spatially. Of course the satellite approach tends to produce vast amounts of data from large tracts of ocean, a very different requirement from that of a marine energy research institute needing to monitor wave conditions at a particular test site.

2.2 Wave energy conversion

2.2.1 Introductory

The progress of WECs depends on a number of crucial factors, including efficient technical performance; economic manufacture, installation and operation; high reliability and survivability in extreme conditions; and acceptable environmental impact. They are certainly not unique to wave devices, but designing for reliability and survivability is especially challenging in one of the harshest energy environments on Earth – the wild, ever-changing, oceans. There is little point in designing a WEC that works well in gentle and moderate seas, only to discover that the first major storm wrecks it. In deep oceans a monster 'hundred year' wave is always a possibility. Reliable operation over, say, 20 years should be the aim, and if that implies reduced efficiency in moderate seas, it is a price that must be paid.

Newcomers to wave energy often find the technical diversity of WECs a little bewildering. The aim may be clear, but there seem to be few overarching principles to guide the effective harnessing of wave energy. Back in Section 1.3.1 we described a number of historical devices, from shoreline 'wave motors' to oscillating water columns, from floating 'ducks' to storage reservoirs fed by tapering channels – a truly eclectic mix. In the early days there was a natural tendency to explore shoreline and near-shore energy conversion before tackling deep-water challenges; today there are plenty of designers willing to chance their reputations offshore. Yet the diversity of approaches remains and it is unclear which will contribute most towards the gigawatt industry hoped for by so many people.

It is natural for researchers and designers to draw on expertise already built up in related industries. By the 1970s a large body of information was available on the hydrodynamics of ships and offshore structures, information that undoubtedly influenced the wave energy community [4]. Yet there were important gaps because WECs must not only survive the waves, they must also generate energy efficiently from them and transfer that energy to shore. Machines placed near the shoreline tend to work in very different wave climates from those well out to sea, and shore-based systems including oscillating water columns have little in common with conventional marine engineering. As the wave energy industry develops and matures it must inevitably strike out in new directions, technical and operational, producing designers and engineers with their own unique set of skills and experience.

2.2.2   Types of wave energy converter

The wide variety of WEC designs has naturally led people to suggest grouping them in categories. Typically, they are classified according to how they work (operating principle); where they work (shoreline, near-shore, offshore) or by their current stage of development (first, second, third generation [4]). Clearly, a comprehensive classification of a particular WEC will include all three components.

The operating principles of WECs may be described in various ways. Here we adopt the system recommended by the European Marine Energy Centre (EMEC) [7], based in the Orkney Islands, Scotland, which covers most devices currently at an advanced stage of development:

- *Attenuator*. This is a long, thin, floating device with its long dimension aligned parallel to the direction of the waves. Typically, it is an articulated, snake-like structure with a number of sections that bend (and perhaps twist) as the waves pass by. It extracts energy from the waves as they pass along it, causing progressive attenuation (reduction in height). A historical precursor is the contouring raft shown in Figure 1.15. There is current interest in flexible attenuator structures which change their shape or volume, compressing fluid as part of the power take-off system.

- *Point absorber*. This type of floating device has horizontal dimensions at the waterline which are small compared with a wavelength. In effect it 'samples' the wave pattern at a particular point. The operation of a point absorber is fairly easy to visualise and it was the first type of WEC to achieve large-scale deployment, notably in the navigation buoys pioneered by Masuda in Japan (see Section 1.3.1).

- *Oscillating wave surge converter*. This device reacts to surging waves and the movement of water particles within them and is aligned at right angles to the oncoming waves. Typically, it has a flap, or arm, pivoted to allow forward and backward movement as each wave passes by. It may be moored offshore, or secured to the sea bed close to shore. The *Nodding Duck* [8] designed by Stephen Salter in the 1970s is the best-known historical precursor of this type of device (see Figure 1.14).

- *Oscillating water column*. This consists of a hollow tube or structure, open to the sea below the water line and enclosing a column of water topped by a column of air. Oncoming waves cause vertical oscillations of the water column, which in turn compress and decompress the air, driving a turbine. Oscillating water columns have a long history (see Section 1.3.1), and a modern installation on the Isle of Islay in Scotland has been feeding electricity into the grid [9] since 2000.

- *Overtopping device*. An overtopping device captures water from waves, directs it into a reservoir above mean sea level, and returns it to the sea via one or more hydroelectric turbines. The device can be on shore (see Figure 1.15) or floating offshore, and may use natural features or specially-shaped collectors or ramps to increase water capture.

- *Submerged pressure differential*. Typically attached to the sea bed below the waves, this type of device uses the pressure differentials caused by the rise and fall of waves to generate electricity.

- *Other*. A few devices do not fall naturally into any of the above categories. They may be unique (and controversial) in their design principles, but it is important to keep an open mind about their potential in the present fast-moving wave energy scene.

In this book we will focus on devices at an advanced stage of development. Most have already reached full-scale prototypes deployed in real-sea conditions, with arrays firmly in prospect, placing them in the *third generation* device category [4]. Figure 2.15 shows examples of the first five above categories, all of which are covered by case studies in Chapter 4.

Our discussion of wave spectra in Section 2.1.3 emphasised that a random, real-life, sea may contain waves of many frequencies, arriving from many directions. Its spectrum is a function of both frequency and direction. The wave rose in Figure 2.9 showed typical variations in wave energy and SWH as a function of azimuth (compass bearing) for a location near a west-facing coast in the north Atlantic. Wave roses for sites in mid-ocean may be much more spread than this; for sites in channels aligned with the prevailing wind direction, or on sloping beaches where refraction tends to align wave fronts with the shoreline, they may be much narrower. All this has big implications for the various types of WEC mentioned above because some are much more directional than others. A point absorber can extract energy from waves coming from any direction. Other types, for example attenuators and oscillating wave surge converters, tend to be highly directional and should be aligned parallel or perpendicular to the wave direction, respectively, if they are to operate efficiently. This may be hard to achieve in random seas. On the whole the best locations for directional devices are those that receive plenty of near-sinusoidal, long-distance swells arriving from a well-defined direction – the extensive

(a) (b) (c)

(d) (e)

Figure 2.15 Examples of five categories of WEC: (a) Attenuator (Pelamis Wave Power Ltd); (b) point absorber (Ocean Power Technologies, Inc.); (c) oscillating wave surge converter (Aquamarine Power Ltd); (d) oscillating water column (Voith Hydro Wavegen Ltd) and (e) overtopping device (Wave Dragon ApS).

west-facing coast of Portugal is a good example. The overall point to be emphasised is that a site's total *wave climate* – direction, strength and variability – is crucial for deciding its suitability for a particular type of WEC.

Another major issue is survival. The various categories of WEC listed above tend to have different robustness. For example, it is probably simpler (though not necessarily cheaper) to design survivability into a shore-based oscillating water column device, encased in concrete and built into rocks or a man-made breakwater, than a near-shore oscillating wave surge converter or an attenuator located several kilometres offshore. Successful WEC design presents a whole range of engineering challenges.

2.2.3 Principles of wave energy capture

2.2.3.1 Floating devices

Some key physical principles underlie the successful capture of wave energy. We start by considering a large subset of WECs that are essentially

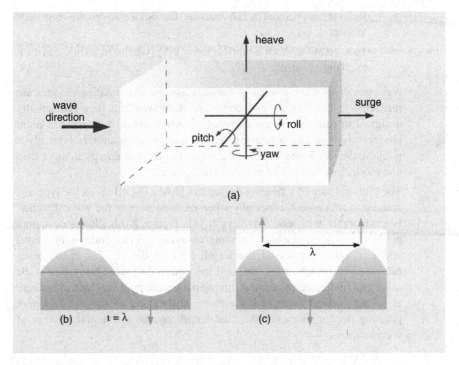

Figure 2.16 Movement of a floating device: (a) translation and rotation terminology; (b) a pure pitching condition and (c) a pure heaving condition.

'floating devices'. Deployed near-shore or offshore, they move in various ways in response to the waves and our first task is to clarify some important terminology which, incidentally, they share with ships and aircraft.

Figure 2.16a shows a simple floating device that is able to move in a variety of ways. Three are *translational*, without rotation:

- *Heave*: vertical up-and-down movement.
- *Surge*: horizontal movement parallel to the wave direction.
- *Sway*: horizontal movement perpendicular to the wave direction.

There are also three types of rotational movement:

- *Pitch*: rotation about a horizontal axis (HA) causing the front and back of the device (in a ship, the bow and stern) to oscillate up and down.

- *Roll*: rotation about a HA causing the device to 'rock' from side to side.
- *Yaw*: rotation about a vertical axis (VA) (which, in a ship, causes it to change course).

Newcomers to wave energy sometimes assume that floating devices are restricted to bobbing up and down on the waves, but they are actually designed to respond to various types of movement. For example, a point absorber is typically a heaving device and an attenuator is typically a pitching device. Some WECs are designed to capture energy in more than one mode, for example by pitching and rolling.

The physical size of a floating device can have a big effect on the type and amount of movement, especially when its dimension in the wave direction is comparable with one wavelength [1]. Figure 2.16b illustrates a pure pitching condition in which a complete wavelength (λ), comprising a crest and a trough, occurs over the length (l) of a device. The crest causes increased buoyancy at one end and the trough decreased buoyancy at the other end (indicated by arrows), giving maximum pitching. However, there is no net vertical force and, therefore, no heaving motion. In fact pure pitching occurs whenever the float length equals an integral number of wavelengths:

$$l = N\lambda, \quad N = 1, 2, 3 \tag{2.12}$$

Figure 2.16c shows a pure heaving condition with one and a half wavelengths occurring within the device length, giving an 'extra' crest. There is now a net vertical force, but no tendency to pitch. In general, heaving occurs whenever the float length is much smaller than a wavelength (in which case it effectively 'samples' the height of each passing wave), or when it equals an odd number of half-wavelengths:

$$l = N\lambda/2, \quad N = 1, 3, 5 \tag{2.13}$$

We see that pitching and heaving of a floating device depend on the ratio between l and λ. Such effects tend to be most significant when it is responding to long-wavelength sinusoidal waves produced by a well-developed ocean swell.

2.2.3.2 Tuning and damping

As waves pass by a floating device they generate cyclic forces and it has long been understood that many devices achieve maximum energy capture when *tuned* to the wave frequency [1, 4]. It is rather like pushing a child on a swing. Give a shove at the right instants (once per cycle of the swing's 'natural' oscillation) and the swing goes higher and higher; but get the timing wrong and you have an unhappy child on your hands! In general a WEC has one or more *natural frequencies* at which it prefers to oscillate. For example, if a point absorber on a calm sea is pushed down slightly and then released, it will bob up and down at a natural frequency determined by its mass and buoyancy. Reduce the mass and/or increase the buoyancy, and the frequency increases. If sinusoidal waves with a frequency close to the natural frequency start to arrive, the device will tend to oscillate enthusiastically. The condition is known as *resonance*, and tuning to achieve it is one of the most important principles of wave energy conversion.

In some devices *fixed tuning* is built in at manufacture to give a good overall match to the wave conditions at the intended location. However, wave frequencies change as seas develop and subside. To accommodate this some devices incorporate *slow tuning* over time scales from many minutes to hours, allowing them to adapt to variations in sea state and maintain resonance. The most sophisticated type of tuning adapts the device to individual waves or small groups of waves but, needless to say, such *fast tuning* is relatively difficult and expensive to achieve.

One important caveat about tuning is that resonance occurs regardless of whether the waves are large or small. Very large waves at resonant frequency may produce damaging forces and movements in exactly the conditions where a device should be limiting its response in order to survive, and special protection measures may be needed. A high degree of tuning has dangers as well as benefits.

There is one further property, apart from mass and buoyancy, which has a major effect on the resonant response and is also crucial to energy capture. This is referred to as *damping*.

The effects of damping can be illustrated by some simple examples and we start with a familiar experiment. Imagine holding the long blade of a kitchen knife firmly down on a table, as in Figure 2.17a, then deflecting the handle with the other hand and letting it go – an experiment that amuses almost every physics student. The knife vibrates up and down at its natural frequency, with diminishing amplitude, and eventually comes to rest. But why do the oscillations decay with time? It is due to damping, caused mainly by air resistance as the knife blade bends backwards and forwards. Without any damping the oscillations would continue indefinitely; but with damping, they decay.

Figure 2.17b shows an air-filled tube floating in deep water, weighted so it stays upright. Push it down slightly then let go and it heaves up and down at its natural frequency, the oscillations decaying with time due to the

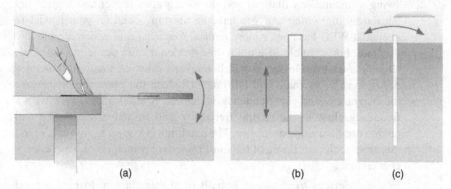

Figure 2.17 Simple oscillating systems: (a) kitchen knife; (b) floating tube and (c) flexible flap.

damping effects of the water. If sinusoidal waves arrive, with a frequency close to the natural frequency, the device resonates and the size of the oscillations is limited by the amount of damping. A third example, shown in Figure 2.17c, represents a flexible flap firmly set into the sea bed. It too has a natural frequency and may display resonance in surging waves. It is risky to stretch analogies too far, but you may have noticed that the floating tube looks rather like a heaving point absorber and the flap looks rather like an oscillating wave surge converter – although we have not so far addressed the question of energy capture.

The classic approach to analysing the effects of damping is to consider a linear system disturbed by a sinusoidal force of variable frequency. Sinusoidal inputs to a linear system have the special property of producing sinusoidal responses at the same frequency, but with variable magnitude (and phase). The simplest type of system to exhibit all the effects we have discussed is of *second-order*, containing mass, stiffness (equivalent to buoyancy) and damping. The magnitude (amplitude) of its response to different frequencies, shown by the *frequency response* in Figure 2.18, depends crucially on the *damping constant* (ξ). Consider, for example the red curve for which $\xi = 0.15$. As the frequency of the sinusoidal input is increased from zero, the response amplitude passes through a resonant peak in the region of the system's natural frequency f_n and then falls away. It is not hard to imagine that strong resonance in a lightly damped WEC can produce oscillations and stresses which may even threaten its survival. It is rather like the opera singer who can shatter a wine glass by producing a loud note at exactly the right frequency. 'Fine tuning' has advantages, but it also involves risk.

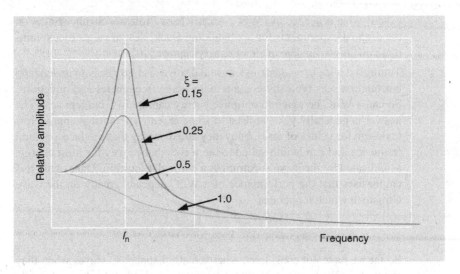

Figure 2.18 Frequency response of a second-order system for different amounts of damping.

As the damping constant increases, the resonance becomes less marked, shown by the orange and yellow curves in the figure. If $\xi = 1$ there is no resonant peak (green curve) and the system is said to be *critically-damped;* if $\xi > 1$ it is *overdamped.* Note that there is a very important trade-off between tuning and *bandwidth;* a highly tuned system responds enthusiastically to a small range of frequencies, but as the damping is increased it responds more modestly to a wider band of frequencies. This has major implications for tuning WECs. Finely-tuned, low-bandwidth, devices do well in long-distance swells characterised by a narrow range of wave frequencies – but may risk damage in high seas. Less highly tuned devices are more suited to random seas with a wider range of wave frequencies – and are generally less vulnerable.

Damping is also an essential part of energy capture. We may go right back to Isaac Newton's insights on the relationship between force and movement to explain this fundamental truth. During a cycle of oscillation the mass in a system stores and releases kinetic energy, stiffness (or buoyancy) stores and releases potential energy, but it is only damping that can *absorb* or *dissipate* energy. Whereas mass resists acceleration and stiffness resists displacement, damping resists *velocity.* Absorption of energy can only occur if force is accompanied by velocity. Damping due to friction or turbulence (as in Figure 2.17) simply dissipates the energy as heat, but damping carefully designed into a system can also generate useful power, exactly what is needed in a WEC. An article by Professor Stephen Salter on the original experiments for his 'Nodding Duck' design elegantly explains these and other fundamentals of energy capture [10].

Tuning a device to resonate in sinusoidal waves is fairly easy to understand but random seas typically contain many wave frequencies and directions. So can a WEC be tuned to optimise energy capture in a random sea? The answer is generally yes – at least to some extent. The normal approach is to design for values of mass, buoyancy and damping that produce a natural frequency and bandwidth suited to the waves that convey the most energy, as indicated by the wave spectrum at a particular location. Once again this emphasises that the performance of a WEC depends greatly on the wave climate in which it operates.

2.2.3.3 When waves meet WECs

So far we have not considered what actually happens to waves when they encounter a WEC or, to put it the other way round, what the device must do to the waves if it is to extract useful energy. A fundamental principle is that absorption of wave energy may be thought of an interference phenomenon [2], summarised by an apparently paradoxical statement:

> To absorb a wave is to generate a wave; to destroy a wave is to create a wave.

This means that a good wave absorber must also be a good wave generator, and vice versa. We illustrate this key idea in Figure 2.19.

Figure 2.19a shows what happens when incident sinusoidal waves meet a smooth, rigid, vertical barrier (sketch 1). They are perfectly reflected (sketch 2) and interfere with the incoming waves to produce a standing wave pattern (sketch 3). No energy is absorbed by the barrier, it is all sent back in the direction from which it came (red and blue arrows). Such perfect reflection has already been discussed in Section 2.1.4 and illustrated by Figure 2.10.

In Figure 2.19b the vertical barrier is replaced by a smooth hinged flap. If the flap is held rigidly vertical (sketch 4) the incident waves are again reflected back (sketch 5, blue curve). If the flap is oscillated backwards and forwards on otherwise calm water it generates its own waves (sketch 5, orange curve). We assume linearity, so these results may be added together (superposed) to give the overall effect – incident waves approaching the oscillating flap and being completely absorbed by it (sketch 6). Note that the flap's effectiveness as an absorber depends on its ability to generate waves with the same frequency as the incident waves, and with the correct *phase* – trough coinciding with crest and vice versa (sketch 5). The

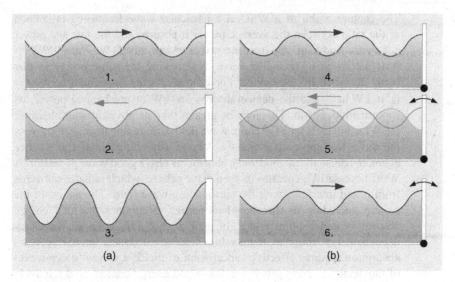

Figure 2.19 (a, b) Reflection and absorption of incident waves (see text for detail).

generated waves travel in the same direction as the reflected waves and cancel them by interference. In a nutshell, the oscillating flap *absorbs a wave by generating a wave.*

But what has happened to the absorbed energy? If we assume the flap is smooth and frictionless, it must have been accepted by the oscillating mechanism and put to some other use. Of course it could be dissipated as heat but if the flap is part of a WEC this is hardly the intended result! No, the absorbed energy must be accepted as mechanical energy and converted to electricity by a suitable generator. For this to happen the motion of the flap must be resisted by a force proportional to instantaneous velocity – the damping effect we mentioned previously. We also know that, according to Newton's third law of motion, to every action there is an equal and opposite reaction, so the wave-generated force must always be opposed by an equal reaction force, supplied from within the device or external to it.

Figure 2.19 neglects friction and turbulence effects and assumes perfect absorption without any onward transmission of the incident wave energy. We cannot expect such idealised performance in practice, although a sophisticated WEC can capture a large proportion of the wave energy it intercepts [10]. Or perhaps we should say *appears to intercept,* because the effective width of a device is generally not the same as the physical width it presents to an approaching wave front. This leads us to one of the most important performance measures of a practical WEC – its *capture width.*

The capture width of a WEC at a particular wave frequency is defined as the ratio between the average power it absorbs and the average power per metre width of the incident waves. Alternatively we may think of it as the width of incoming wave front having the same power as that being absorbed by the device. For example, if the incident wave power is $40 \, \text{kW} \, \text{m}^{-1}$ and the device absorbs $360 \, \text{kW}$ of mechanical power, its capture width is $9 \, \text{m}$. This may be greater than its physical width because the WEC interacts with the entire wave field surrounding it and not just with the waves that approach it directly. A good way of explaining this is, once again, to realise that a good wave absorber is also a good wave generator. A WEC necessarily generates its own wave pattern, which radiates outwards from it and interferes with the surrounding wave field. The nature of the pattern, including its directional properties, determines the total energy absorption and capture width. In general, both are strongly frequency dependent, reaching maximum values close to resonance. Intuitively, good absorption requires effective cancellation of incident waves 'down-wave' of the device, reducing their amplitude and energy content, and relatively little energy loss by radiation 'up-wave' or to the sides. Unlike the two-dimensional simplicity of Figure 2.19, interference effects are essentially three-dimensional.

A good example of a device with a capture width much greater than its physical width is a long, snake-like, attenuator – see, for example Figure 2.15a. A small physical width is presented to the incident wave front, but as waves pass along its length there is continuing absorption of energy due to interaction with the surrounding wave field.

An effective WEC must clearly be a good absorber of wave energy, but it is every bit as important that it survives storm conditions – one of the greatest challenges facing design engineers. Discussions of the interactions between a WEC and the waves that power it, illustrated by linear models and assuming low to medium wave heights, risk losing sight of the huge forces generated by extreme seas and the paramount need to protect against them. In such situations there is a dangerous surfeit of energy and efficiency of capture becomes irrelevant.

We end this section with a few comments about another threat to WECs, the process known as *fatigue*. This is essentially an ongoing, long-term, problem rather than a short-term one posed by a vicious storm. It occurs generally in engineering materials subjected to repeated cyclic loads and eventually – perhaps after millions of cycles or oscillations – causes failure. Typically, small cracks start to develop in the material or component in the region of maximum stress and grow with repeated cycles. Susceptibility to fatigue depends very much on the material: try bending a thin strip of aluminum sheet backwards and forwards, and it soon fails; mild steel survives much better. Today's WECs contain a variety of structural materials, principally steel but also materials such as glass reinforced plastic (GRP). There is also increasing interest in flexible materials for use in devices that change shape or volume as part of the power take-off mechanism. All have their own distinctive fatigue characteristics.

Development of fatigue depends on the magnitude as well as the number of load cycles, making resonant conditions with low damping especially risky. Fatigue is also hastened when a component or structure, already stressed by a steady load, has a cyclic load superimposed on it – a common situation in WECs. Progressive fatigue is to be expected in oscillating devices and should be considered a normal part of wear and tear. But a large WEC must also cope with rogue waves and storms that force its control and protection systems into rapid response. When discussing the slow and steady onset of fatigue, we should remember that fatigue failure may be hastened by extreme events that occur randomly during a WEC's lifetime. Survival may not be a 'principle' of wave energy capture, but it is certainly a prerequisite.

Our discussion of energy capture has focussed on fundamentals – types of movement, resonance and damping, frequency response, wave absorption and generation, capture width – which apply, to a greater or lesser extent, to most types of WEC. There are, however, a few designs with special features needing additional explanation. Foremost among these in today's wave energy scene are oscillating water column and overtopping devices, and we will meet important examples in Chapter 4.

2.3 Tidal streams

2.3.1 Hydrodynamics

Standing on a shoreline or promontory, watching the tides ebb and flow, it is tempting to imagine that the strength and direction of the tidal stream is well-behaved and easily predicted. But, as Chapter 1 has shown, the influences on tides are many and varied, from the ever-changing gravitational pull of the moon and sun to the complexities of coastal geography. It may sound simple to place a turbine in a tidal stream and tap its energy, but in practice tidal flows are complicated and site-specific. Design engineers must understand their peculiarities if they are to have any chance of predicting device performance and maximising energy capture. The design of turbines requires an intimate knowledge of how water flows over rotor blades and generates useful torque. So we devote this section to some basic aspects of *hydrodynamics* – the science of fluids in motion which lies at the heart of tidal stream technology [11].

(a)   (b)

Figure 2.20 (a) Daniel Bernoulli and (b) Theodore von Kármán (Wikipedia).

One of the most basic principles in hydrodynamics is due to the Dutch–Swiss mathematician Daniel Bernoulli (1700–1782) (Figure 2.20a) who showed that, in a steady flow, the sum of all forms of mechanical energy in a fluid remains constant along a streamline. This means that any increase in speed must be accompanied by a reduction in pressure and/or potential energy. Bernoulli's equation, which derives from the conservation of energy, is most readily applied to incompressible liquids. However one of the simplest ways of demonstrating the idea is illustrated in Figure 2.21a which shows air flow being measured by a *venturi meter*. As the air passes from the wide to the narrow section of the tube its velocity increases from u_1 to u_2, so the pressure must reduce. This is indicated by the difference in height h between the two liquid columns (in this case the streamline is horizontal so there is no change in potential energy).

Assuming water to be incompressible, with negligible viscosity, Bernoulli's equation takes the form:

$$p/\rho + gz + u^2/2 = \text{constant} \qquad (2.14)$$

Here p, z and u denote the water pressure, height and velocity, respectively; ρ is the density of water and g is the gravitational constant. In effect

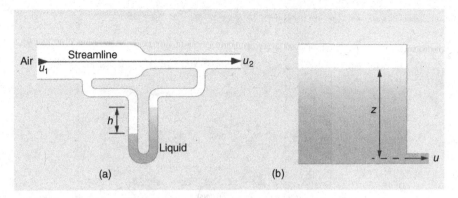

Figure 2.21 (a, b) Applications of Bernoulli's principle (see text for detail).

the equation states that the total mechanical energy, made up of potential energy due to pressure and height, plus kinetic energy due to velocity, remains constant at all points along a streamline.

One practical implication of Bernoulli's equation is that, when water flows out of a reservoir, the sum of all forms of energy is the same along all streamlines because the potential energy per unit volume of water is the same everywhere in the reservoir. A limiting case arises when we wish to calculate the maximum possible drain rate of a tank (or reservoir) with a hole or pipe at its base, illustrated in Figure 2.21b. The total energy of water at the top of the tank is entirely potential, due to the height z, but as water exits the tank with speed u its energy is entirely kinetic. Potential energy is being converted to an equivalent amount of kinetic energy. Equation 2.14 shows that:

$$u^2/2 = gz \quad \text{and therefore} \quad u = (2gz)^{1/2} \tag{2.15}$$

For example, if a tank is filled to a depth of 5 m, the exit speed of water at the base is $(2 \times 9.81 \times 5)^{1/2} = 9.9 \text{ m s}^{-1}$. In practice the speed will be slightly reduced by viscosity and friction.

Equation 2.14 takes no account of important frictional effects that arise in tidal engineering, including turbulence in tidal streams and the reduction of flow speeds close to a sea bed. Almost two centuries after Bernoulli's pioneering work was published another famous scientist, a Hungarian-American by the name of Theodore von Kármán (Figure 2.20b), added his own major contributions to the science of fluid flow. Known principally for his work in aeronautical engineering, von Kármán proposed the following

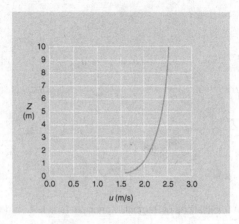

Figure 2.22 Speed profile of a tidal stream close to the sea bed.

equation to describe the flow speed u of a turbulent fluid near a horizontal boundary:

$$u = u_* / k \ \ln z/z_0 \tag{2.16}$$

where u_* is known as the *shear velocity* or *friction velocity*, k is the *von Kármán constant*, z is the height above the boundary and z_0 is the *roughness height*. Applied to flow in a tidal channel, von Kármán's equation predicts a logarithmic relationship between the water speed and the height above the sea bed.

It is important to realise that the speed reduction near the boundary is caused by *shear stresses* set up by the bed's roughness or *drag*. The roughness height is the height at which the speed theoretically reduces to zero and is around one tenth of the physical height of surface roughness elements such as gravel or sand ripples. Since the bed is assumed to be 'non-slip', reducing the flow speed to zero at the boundary, von Kármán's equation is often referred to as 'the law of the wall'. It has been found to agree closely with speed profiles actually measured in the sea and is important for tidal turbines placed on the sea bed. As an example Figure 2.22 shows the speed profile predicted for $u_* = 1 \ \mathrm{m \ s^{-1}}$, $z_0 = 0.3$ mm and $k = 0.41$ (the generally accepted value), relevant to a seabed mix of mud, sand and gravel [11]. In this case the flow speed rises exponentially from about 2.0 to 2.5 $\mathrm{m \ s^{-1}}$ as the height above the sea bed increases from 1 to 10 m – a range of great importance for tidal turbines. A whole family of speed profiles can be generated, depending on the values of friction velocity and roughness height.

(a)   (b)

Figure 2.23 Osborne Reynolds (a) and his famous experiment (b) (Wikipedia).

The work of von Kármán developed out of some fundamental insights into turbulent flow gained by scientists and engineers during the Victorian period. Eminent among them was Osborne Reynolds (1842–1912), a mathematics graduate of Cambridge University who later became a professor of engineering in Manchester. He was particularly interested in the effects of speed on flow patterns close to boundaries. Basically, low-speed flow tends to be smooth or *laminar*, but above a certain speed it becomes increasingly chaotic. In a famous experiment in 1887 (see Figure 2.23) Reynolds forced water from a reservoir into a long glass tube and injected a streak of dye. As the speed of the water increased, the streak of dye changed from a straight line into a complicated series of vortices and lateral fluctuations, indicating turbulence. Such effects have great practical significance for oil and water flow in pipelines, air flow over the wings of aircraft and turbulence in tidal streams. They depend on the value of an extremely important dimensionless quantity known as the *Reynolds number*.

The Reynolds number is a measure of the relative importance of inertial and viscous forces acting on a fluid in motion. Inertial forces depend on mass or density, whereas viscous forces depend on viscosity or 'stickiness'. Treacle and water may have comparable densities but treacle is far more viscous and a body moving through it experiences much more drag. The Reynolds number (Re) is defined as:

$$Re = UL/v \tag{2.17}$$

where U is the mean flow speed, L is a linear dimension of the system and ν is the *kinematic viscosity*. A low Reynolds number implies a slow-moving, sticky fluid in which viscous forces dominate and the flow is streamlined or laminar; a high Reynolds number implies a relatively fast-moving, non-viscous fluid dominated by inertial forces which produce turbulence. Intermediate values – typically between 500 and 2000 – fall in a transition region between laminar and turbulent flow. Tidal streams are normally turbulent – the condition modelled by von Kármán's equation – apart from an extremely thin laminar layer adjacent to the sea bed.

Turbulence is important in practice because it affects the performance of tidal turbines and places unwelcome stresses on their rotors. The velocity of a tidal stream can be considered as composed of a horizontal component which produces the basic driving force for the turbine, together with random fluctuations in the horizontal, vertical and lateral directions caused by turbulence. Vertical and lateral components are of no use for energy production. The scales of turbulence vary greatly, from small eddies generated near the sea bed to very large ones produced by lateral shear against coastlines or headlands. The various eddies interact strongly, tending to break down to scales where their energy is dissipated as heat by viscosity.

Apart from their indication of laminar or turbulent conditions in a stream of fluid, Reynolds numbers are critically important for testing model turbines in a water tank. In general, it is not possible to transfer measurements made on a small-scale model to a full-scale prototype because they are not *dynamically similar*. However, they become so if the Reynolds numbers can be adjusted to similar values in the two situations. In this sense they provide the essential theoretical link between laboratory testing and full-scale operation. Another dimensionless number that is important in tank testing is the *Froude number*, which is more relevant to WECs than tidal stream turbines. We shall return to this topic in Section 2.5.1.

2.3.2 Tidal harmonics

The basic hydrodynamics outlined in the previous section has a huge range of applications in practical engineering and is by no means restricted to marine energy. We now move on to a topic with a fascinating history that also has a special relevance to tidal stream engineering – the cyclic, or *harmonic*, nature of tidal patterns that is responsible for the highly variable outputs from tidal turbines [11]. We have already discussed basic aspects of the world's tides in Section 1.2.2 and it may be helpful to recall a few key points before proceeding:

- The rise and fall of tides in a particular location depends on the ever-changing gravitational attractions of the sun and moon, the earth's rotation, the movement of huge quantities of water around continents and islands, and complex features of coastal geography and bathymetry.

- The strength of a tidal stream, which flows back and forth in sympathy with the local tides, is greatly affected by the shape of the channel and the sea bed. Strong tidal streams do not necessarily occur in locations displaying large tidal ranges.

The oscillations of tidal streams have a major influence on the efficiency, capacity factor and practical operation of tidal turbines and other devices, and are directly related to the special features of local tides. An appreciation of tidal phenomena is therefore central to tidal stream technology.

The World's tides have fascinated philosophers since ancient times, but a true understanding only began to develop as part of the extraordinary blossoming of the natural sciences in seventeenth century Europe [11]. We focus here on three historical giants of physics who, among their many and varied interests, devoted time to tidal science; and one twentieth century scientist, largely unknown to the general public, who made a remarkable contribution to his chosen specialism (Figure 2.24).

Isaac Newton started the modern era of tidal science by showing for the first time that tide-generating forces are due to the attractions of the moon and sun. His universal theory of gravitation, published in the *Principia* of 1687, explains the spring–neap cycle and why there are generally two tides in a day rather than one. Newton's revolutionary approach focussed on tides that would occur in an ocean evenly spread over the whole surface of the earth, without attempting to include effects of land masses and bathymetry. In this sense it was only a partial theory, but it moved tidal science a giant leap forward and held out the promise of future advances.

(a)　　　　(b)　　　　(c)　　　　(d)

Figure 2.24 Pioneers of tidal science: (a) Newton, (b) Laplace, (c) Kelvin and (d) Doodson (Wikipedia).

Newton's explanation of the tides relied on his three laws of motion as well as the law of gravitational attraction:

- *First law of motion*: every object remains in a state of uniform motion unless disturbed by an external force.

- *Second law of motion*: the applied force is equal to the mass of the body multiplied by its acceleration.

- *Third law of motion*: action and reaction are equal and opposite.

- *Law of gravitation*: the gravitational attraction between two bodies is proportional to the product of their masses and inversely proportional to the square of their distance apart.

These laws illuminated a vast range of phenomena in physics and astronomy as well as underpinning what is now known as *Newton's equilibrium theory of the tides*. Its main limitation was the inability to predict tidal times and heights in a particular place due to the complexities of coastal geography and bathymetry. The Royal Society of London, founded in 1660 and subsequently to appoint Isaac Newton as its president, was keen to back up new theories with practical experimentation and encouraged the installation of tide gauges to record tidal ebbs and flows. As the years went by there was also increasing interest in the complicated spatial and temporal variation of tides along coastlines, an extremely important issue for naval and merchant shipping. The complex coastal geography of Britain, linked to huge volumes of water in the Atlantic Ocean, English and Irish Channels, and North Sea, produce large tidal variations over comparatively short distances. For example, maximum tidal ranges vary between 14 m in the Bristol Channel and only 2 m near Southampton; high water (HW) at Dover coincides with low water at the western end of the English Channel, and so on. Such details could only be established by careful observation over long periods.

The next major historical figure to advance tidal science was the French mathematician and astronomer Pierre Simon de Laplace (1749–1827) who, almost exactly 100 years after Newton's *Principia*, published his *tidal equations*. The three partial differential equations, which formed the first-ever theory of tidal dynamics, relate the ocean's horizontal flow to its surface height and are used for tidal computations to this day. They explain three major types of tidal pattern:

- *Semi-diurnal*: the familiar pattern of two high and two low tides per day.

- *Diurnal*: the less common pattern of one high and one low tide per day.

- *Mixed*: a combination of the two.

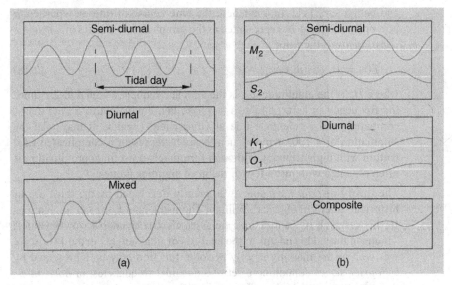

Figure 2.25 (a) Major tidal patterns and (b) harmonic constituents.

Typical plots of height against time for the three types are shown in Figure 2.25a. There is normally some difference between the heights of successive tides in semi-diurnal and diurnal patterns, although the details vary greatly over time and between locations; mixed tides display more complex irregularities. The prevalence of a particular type depends crucially on the shape of the ocean basin. If the earth were covered by a uniform ocean without any continents, two equal high and low tides per day would occur everywhere. But continents and land masses tend to block the westward progress of tidal 'bulges' as the earth rotates, disturbing tidal patterns. Conditions in adjacent basins may vary greatly, the classic case being the Bay of Mexico where the tides are largely diurnal, contrasting with the semi-diurnal tides along the eastern seaboard of the USA. Diurnal tides also occur in parts of East Asia; Western Europe is semi-diurnal and the Pacific coast of the USA is well known for its mixed tides. All these special features and peculiarities affect the characteristics of local tidal streams, including peak velocities and the timing of ebb and flow phases.

Laplace realised that repetitive features in a tidal pattern must correspond to the periodicities, or frequencies, present in the tide-generating force that produces them. We now know that subtleties in the relative movements of the earth, moon and sun produce a surprisingly large number of such frequencies and that the tidal pattern at a particular location represents the superposition of its responses to them all, modified by coastal geography and bathymetry [11]. In a nutshell, the time course of tides at a particular port or location may be represented as the summation of a series of *harmonic constituents* of the form:

$$f(t) = H_n \cos(\omega_n t - \varphi_n) \tag{2.18}$$

where H_n is the amplitude; ω_n is the angular frequency (equal to 2π times the frequency in hertz) and φ_n is a phase lag, equivalent to a time delay, compared with a reference tide at Greenwich, London. It may take the summation of a 100 or more such terms to represent a complicated tidal pattern with high accuracy, although fortunately a useful approximation is often obtained with just a few of the most significant ones.

The theory of *tidal harmonics* was greatly developed by the third of our historical 'giants of physics', William Thomson, later Lord Kelvin. Famous for his research in such diverse fields as electricity and thermodynamics, Thomson turned his talents to the analysis of tidal heights in the 1860s. He was well aware that any strictly periodic function may be represented by a *Fourier series* containing a set of sinusoidal frequencies, or *harmonics*, that are integer multiples of the lowest frequency present, known as the *fundamental*. In general, the more harmonics that are included, the better the representation. In the case of tides, however, the situation is much more complicated because tidal patterns in a particular location, although showing repetitive features over various time scales, are not strictly periodic and never repeat themselves exactly. They contain many fundamental frequencies corresponding to the ever-changing relative motions of the earth, moon and sun, so analysis cannot be limited to the harmonics of a single frequency. Any attempt to restrict it in this way would require a huge number of terms and severely limit the time range over which it would apply.

We may appreciate the magnitude of the problem tackled by Thomson if we first consider the state of tidal prediction at that time. By the 1870s the fundamental frequencies present in the astronomical tide-generating force were well understood and could be accurately predicted. The response to each frequency, in amplitude and phase, could be obtained for a particular port or coastal location by careful tide-gauge measurements. So a future tide could, in principle, be predicted by summing all the harmonic constituents, obtained by combining each frequency component in the tide-generating force with the relevant local data. However up to the 1870s all the calculations had to be done by hand using paper, pencil and tables – an immensely time-consuming and tedious business.

William Thomson realised that what was needed was a machine – effectively a mechanical analogue computer – that could calculate the sum of many harmonic constituents with different values of amplitude, frequency and phase, over a considerable time range in the future. His solution,

Figure 2.26 The Thomson tidal prediction machine of 1872–3 (Wikipedia/William M. Connolley).

the beautiful tidal prediction machine of 1872–3, shown in Figure 2.26, is now held in London's Science Museum. It had first to be set with local data and then summed eight harmonic constituents as the mechanism was turned, printing out a prediction of tidal height against time on a moving band of paper. A year's tidal predictions for a given location or port took several hours to produce. Subsequent models were built to accommodate up to 24 constituents. Various designs of mechanical tide-predicting machines were also produced in the USA, Germany and Norway, and some became of strategic military importance in the two World Wars. But by the 1970s they had become museum pieces, their role taken over by that modern workhorse of automatic calculation – the digital computer.

The number of harmonic constituents needed to produce accurate predictions depends on the complexity of tides in a particular location. Simple tidal patterns can often be predicted adequately by summing just four constituents; but up to 100 or more may be needed in some cases. Returning to Figure 2.25, typical curves for the four constituents which tend to dominate are shown in Figure 2.25b, together with a composite curve. Two are semi-diurnal (M_2 and S_2) and two are diurnal (K_1 and O_1); their letters were allocated by another Victorian who made big contributions to tidal science – George Darwin (1845–1912), second son of the naturalist Charles Darwin. Further details of the constituents are given in the table.

Letter	Name	Caused by	Period (h)
M_2	Principal lunar semi-diurnal	Earth's rotation with respect to Moon	12.42
S_2	Principal solar semi-diurnal	Earth's rotation with respect to Sun	12
K_1	Lunisolar diurnal	Moon's declination (together with O_1)	23.93
O_1	Principal lunar diurnal	(see K_1)	25.82

The tides in a particular location may be classified according to the relative amplitudes of these four constituents using a *Formzahl number* defined as:

$$Fz = (K_1 + O_1)/(M_2 + O_2) \qquad (2.19)$$

If *Fz < 0.25* the tides are classed as *semi-diurnal;* if *0.25 < Fz < 1.5* they are *mixed semi-diurnal dominant;* if *1.5 < Fz < 3.0* they are *mixed diurnal dominant;* and if *Fz > 3.0* they are *diurnal.* As examples the next table gives typical amplitudes of the four major constituents at three locations in the United Kingdom and one in the USA, together with Formzahl numbers and classifications:

Location	M_2	S_2	K_1	O_1	Formzahl number	Formzahl classification
Bristol channel	4.3	1.5	0.07	0.08	0.026	Semi-diurnal
Dover	2.3	0.71	0.05	0.06	0.037	Semi-diurnal
Islay	0.16	0.14	0.09	0.08	0.57	Mixed semi-diurnal dominant
Galveston, Texas	0.23	0.071	0.48	0.45	3.1	Diurnal

The first two locations, the Bristol Channel (Avonmouth) and the port of Dover, are emphatically semi-diurnal, although the reduced amplitudes of M_2 and S_2 at Dover reflect its far lower tidal range. The Hebridean island of Islay is an interesting case because the waters off its southwest coast have a very small tidal range by UK standards yet display some of the fastest tidal streams (see Section 1.2.2 and Figure 1.11); and, unusually for the UK, the tides tend to be mixed rather than strongly semi-diurnal, suggesting powerful interactions between waters of the Irish Sea and North Atlantic in the channel between Scotland and Northern Ireland. Galveston, Texas, the port of Houston, is quite different with relatively large K_1 and O_1 constituents, signifying diurnal tides typical of the Bay of Mexico. These four locations are just a few of the many worldwide that we could have chosen to illustrate the great variety of tidal patterns.

Finally, we come to the fourth of our historical contributors to tidal science shown in Figure 2.24 – Arthur T. Doodson (1890–1968). Born in Lancashire, England, and educated at Liverpool University, Doodson produced his own design for a tide-prediction system, the *Doodson–Légé* machine of 1948–9 which accommodated up to 42 harmonic constituents. However, he is best known for an extremely painstaking development of harmonic analysis, and for the seminal paper he wrote in 1921 [12]. This included his invention of the *Doodson numbers* to arrange many hundreds of harmonic constituents in logical groups, a classification which is still in widespread use today.

As we have already seen, harmonic analysis of the tides is not restricted to the harmonics of a single fundamental frequency. In many locations a large number of fundamentals and associated harmonics, modified in amplitude and phase by local geography and bathymetry, must be summed to produce accurate tidal predictions; and the story of prediction machines is of improving accuracy by including more and more constituents. Doodson's major contribution was to reduce complexity by using just six cleverly chosen fundamental frequencies relating to principal features of the moon's orbit around the earth and the earth's orbit around the sun, with periods ranging from 1 day to 20 940 years [13]. He then organised no less than 399 tidal constituents into well-defined groups based on these frequencies, labelling them systematically with Doodson numbers. Few scientists have devoted so much thought over a lifetime's work to understanding the world's tides.

This brief historical review of developments in tidal science has concentrated on the heights of tides, but we must now turn to the interests and concerns of those wishing to install tidal turbines in fast-moving coastal waters – the prediction of tidal streams.

2.3.3   Predicting tidal streams

Our discussion of tides and tidal prediction has so far concentrated on variations in the heights of tides, not their flow rates. Heights are relatively simple to measure and are obvious to every visitor to the sea coast; they

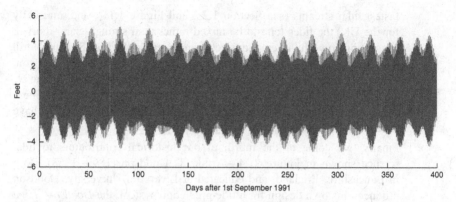

Figure 2.27 Tidal height pattern over a 400-day period at Bridgeport, USA (Wikipedia).

apply over considerable distances simultaneously, or with only a modest time shift. But tidal streams tend to be far more localised, are more difficult to measure, and often change dramatically from one kilometre to the next as water is forced through channels, meets headlands, or enters bays and estuaries. Back in Section 1.2.2 we mentioned many of the factors influencing tidal streams and the regions of the world where they are most powerful. Figure 1.9 illustrated changes in flow and ebb rates over the spring–neap cycle of a semi-diurnal tide – even though such a simple, well-behaved, pattern is unlikely to be found in many locations.

Figure 2.27 suggests why more complicated tidal stream patterns must often be expected in practice. It shows a record of tidal heights (not currents) over a 400-day period at Bridgeport, Connecticut, USA. The tides are basically semi-diurnal, with substantial spring–neap cycles, but the pattern is far from regular – note, for example the big differences between around 50 and 200 days. Such variations are by no means exceptional, and in many locations are accompanied by big changes in tidal flows throughout the year.

Internationally, one of the tidal streams best known for its waywardness occurs in the Cook Strait between the North and South Islands of New Zealand. Tidal heights at the two ends of the strait are almost exactly out of step, high tide at one end coinciding with low tide at the other due to the principal lunar semi-diurnal constituent (M_2) being out of phase. This produces vigorous current surges, but a negligible tidal range in the centre of the strait. The duration of the surges is hard to predict, especially in rough weather conditions, and on occasions the surge in one direction is completely suppressed. The situation is complicated even further because monthly variations are quite different on the north and south sides of the strait. All this must have tried Captain James Cook sorely on his famous voyage of discovery in 1769, and one wonders what New Zealand's tidal stream engineers will make of it in the future.

On the face of it, accurate prediction of tidal streams looks like a big challenge. It is fairly easy to appreciate how tidal bulges, modified by land masses as they progress across the oceans, produce changes in tidal heights and times; but tidal streams, fashioned by the details of local geography and bathymetry, seem a step further away from cause and effect. Relationships between tidal heights and flows are very complex and successful prediction of one does not necessarily aid prediction of the other. Yet tidal currents are created by the same astronomical forces as tidal heights, so logically it must be possible to analyse them using similar methods. Indeed it turns out that the historic progress of tide prediction based on harmonic analysis and tidal constituents, developed by Thomson, Doodson and others, can indeed be extended to estimate the velocities of tidal streams and the times of maximum flow and slack water. Of course this assumes that the necessary observational data have been collected at locations of interest. Although tidal streams are less well documented than tidal heights, the USA, UK and several other countries have made great strides in recent decades and now provide valuable prediction services [14, 15].

In the UK the large-scale nautical charts issued by the British Admiralty have for many years included tidal flow data in the form of *tidal diamonds*, each consisting of a reference letter marking a location on the chart and an associated table [11, 15]. Figure 2.28 shows typical tidal diamond information, in this case for three locations (A, B, C) at the eastern end of

		A 50 42'.3N 0 26'.5E		B 50 53'.0N 1 00'.0E		C 51 01'.0N 1 10'.0E	
Hours	Dir	Sp	Np	Dir	Sp Np	Dir	Sp Np
Before HW 6	248	0.8	0.4	213	1.6 0.9	224	0.9 0.5
5	067	0.5	0.3	214	2.1 1.2	239	1.0 0.6
4	068	1.9	1.0	215	1.8 1.1	235	1.1 0.6
3	071	2.6	1.5	213	0.9 0.5	242	0.6 0.4
2	069	2.3	1.3	Slack		Slack	
1	068	1.2	0.6	033	0.8 0.5	052	0.6 0.3
HW	067	0.1	0.1	032	1.5 0.8	049	1.2 0.7
After HW 1	248	0.9	0.5	031	1.9 1.1	049	1.3 0.7
2	247	1.4	0.8	030	1.7 1.0	056	1.0 0.5
3	251	1.8	1.0	031	1.2 0.6	054	0.5 0.3
4	253	1.7	1.0	032	0.4 0.2	Slack	
5	250	1.6	0.9	211	0.4 0.2	219	0.4 0.2
6	249	1.2	0.7	212	1.3 0.7	217	0.8 0.4

Figure 2.28 Tidal diamond data (Wikipedia).

the English Channel. At each location the flow speeds in knots (1 knot = 0.51 m s^{-1}) and direction are given at hourly intervals before and after *HW*, for average spring (*Sp*) and neap (*Np*) tides. For example, at location *A* the flow rate 3 hours before HW reaches 2.6 knots at spring tides, but only 1.5 knots at neaps; 3 hours after HW the reversed flow is 1.8 knots at springs and 1.0 knots at neaps. It is quite common for flow rates to be greater on the flood phase than on the ebb. Such tidal diamond information is a valuable resource for navigation, although the locations of Admiralty diamonds do not generally match those most suitable for tidal turbines and separate observations must be made. Note also that the tidal diamonds refer to fixed times before and after HW and do not give the peak flow rates which are important for estimating the maximum power intercepted by a turbine.

One recent, and notable, approach to predicting tidal streams combines tidal diamond information with classic harmonic analysis [11]. Current amplitudes for six major harmonic constituents (M_2, S_2, K_1, O_1, as discussed previously, plus M_4 and K_2 to account for shallow-water and lunisolar semi-diurnal effects) are used as data for a computer program, the *simplified tidal economic model* (*STEM*), which synthesises a 12-month time series of current velocities and predicts the energy yield for a specified set of tidal turbines.

If tidal current data are not available for a particular location it is possible, in principle, to perform a *computational fluid dynamics* (*CFD*) analysis [11] to take account of detailed coastal geography and bathymetry. In practice, however, direct on-site measurement of current speeds, directions and depth profiles is almost always the best approach, with one important caveat – weather conditions, especially atmospheric pressure and strong winds, may affect tidal streams substantially and should always be taken into account.

2.4 Tidal stream energy conversion

2.4.1 Introductory

Reliability and survivability are key issues for all machines exposed to tough and corrosive marine environments, and tidal stream turbines and devices are no exception. To be effective tidal stream devices must generate electricity reliably and efficiently over periods ranging from a single tidal cycle to many years. They encounter:

- Complex flow patterns which accelerate from rest to velocities that may exceed 3 m s^{-1} on both flood and ebb tides.

- Energy levels that vary greatly over the lunar cycle, normally peaking at spring tides, and even further around the equinoxes in March and September.

- Short-term fluctuations caused by turbulence.

In principle, tidal stream devices may take a number of forms. Most common in practice are horizontal-axis (HA) turbines with the water flow parallel to the axis of rotation (*axial flow*), which are conceptually similar to the large horizontal-axis wind turbines (HAWTs) that dominate today's wind energy market. There is also interest in vertical-axis (VA) turbines with the flow perpendicular to the axis of rotation (*crossflow*). Another category of device is based on *oscillating hydrofoils* with a reciprocating motion. Before we focus on HA turbines, which are currently seeing a great deal of large-scale development leading towards commercialisation, we should say a few words about the alternatives.

VA turbines have a long history in the wind energy industry so it is natural for engineers to consider similar designs for use in tidal streams. It must be admitted that the present dominance of HA machines in the wind turbine market suggests they have particular advantages and that, as so often in practical engineering, 'the proof of the pudding is in the eating'. The most-quoted advantage of a VA device is that a gearbox and electrical generator may be placed either above or below the rotor, offering design flexibility and potentially simplifying access and maintenance. Over the years many VA designs have been proposed and some have progressed to prototype stage [11]. Two of the best known are illustrated in Figure 2.29:

- *Savonius turbine.* Invented by the Finnish engineer Sigurd Savonius in 1922, the flow into this turbine catches each 'scoop' in turn but slips past its rear surface. In the figure, the turbine is shown with an open top to reveal its characteristic 'S' shape. Although inherently inefficient compared with modern blade turbines, the Savonius, or variations on it, has received occasional attention for tidal stream application.

- *Darrieus turbine.* This type of turbine can claim to be the best researched, and most efficient, of all VA machines, with efficiencies up to about 35%. Patented by the French aeronautical engineer Georges Darrieus in 1927, its wind-energy examples are popularly known as 'eggbeaters'. Based on curved blades with airfoil or hydrofoil sections, the Darrieus is more efficient than the Savonius but tends to suffer from self-generated turbulence, blade vibration and fluctuating torque. The latter problem may be reduced by shaping the blades in the form of a helix. Darrieus rotors do not normally self-start and must be driven up to about twice the speed of the flow before

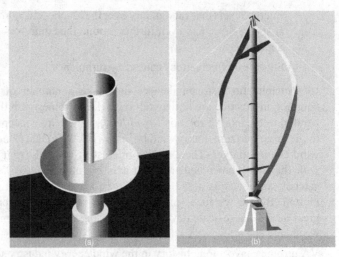

Figure 2.29 Two well-known types of vertical-axis turbine: (a) Savonius and (b) Darrieus.

they produce net torque; they can only be stopped in a strong flow by brakes or spoilers. The vertical axis must either be extremely robust, or supported by some form of guying. A number of variations of the Darrieus have been proposed and investigated, including devices with straight vertical blades attached to horizontal cross arms, but their use in tidal streams seems threatened by problems of reliability and fatigue.

Oscillating hydrofoils offer a very different form of contender for use in tidal streams. They generate power by reciprocating up and down (or backwards and forwards) across the stream rather than by rotating. The hydrofoils must rapidly reverse their direction of travel at the end of each stroke, producing large forces and interrupting the smooth generation of power. Power take-off is also more complicated than in a continuously rotating machine. As an alternative a number of hydrofoils may be attached to a continuously moving horizontal chain. Oscillating hydrofoil devices may be preferred for shallow waters because they can sweep wide rectangular flow areas.

All things considered, it seems likely that HA tidal turbines will dominate the scene in the coming years, not least because the technology has so much in common with wind turbines. From now on we will concentrate on them, making occasional reference to other types of design where appropriate.

2.4.2 Tidal stream turbines

2.4.2.1 Turbine sizes and power ratings

Our discussion of power, energy and performance in Section 1.4 considered the amount of electricity generated by a typical 1 MW tidal turbine over the course of a year and related it to average household consumption. There are a few more general aspects of performance to be covered before we explore the various components of modern tidal turbines – rotors and blades, electrical generators, power electronics and control systems – and explain their operating principles.

A key issue is the amount of power intercepted by a turbine placed in a tidal stream. A well-known equation of hydrodynamics gives this as:

$$P = 0.5 \, \rho \, A \, U^3 \tag{2.20}$$

where ρ is the density of sea water (about 1030 kg m^{-3}); A the area of the rotor facing the stream, often referred to as its *swept area*; and U the stream velocity. For example, the equation predicts that a turbine in a tidal stream flowing at 1, 2 and 3 m s^{-1} intercepts 0.5, 4.1 and 13.9 kW, respectively, for each square metre of swept area. Note how power levels increase dramatically with velocity, and that we have used the word 'intercepted' rather than 'captured' because the figures relate to the power in the stream, not the amount actually extracted by the rotor.

It is hardly an exaggeration to say that the cubic relationship between power and velocity in Equation 2.20 is the single most important technical factor affecting tidal turbine design and performance, for the following reasons:

- A turbine has to operate over a huge range of stream power levels. For example, it may start generating when the stream reaches 0.5 m s^{-1} and reach its rated (peak) output at 2.5 m s^{-1}. A 5 : 1 velocity range corresponds to a 125 : 1 range of intercepted power that must be converted efficiently into electricity.

- The cubic relationship means that short periods of fast-moving stream in each tidal cycle contribute disproportionately to the total energy harnessed by the turbine.

- The turbine must withstand occasional turbulence and surges which substantially exceed the rated velocity.

The proportion of hydraulic power converted into useful mechanical power is denoted by the rotor's *power coefficient*, C_p; and the proportion of mechanical power converted into electrical power by an efficiency figure η.

The overall efficiency of the turbine is therefore $C_p \eta$ and the electrical power output is given by:

$$P_e = 0.5 \, C_p \, \eta \, \rho \, A \, U^3 \tag{2.21}$$

If the rotor of a HA turbine has diameter d, the swept area is $\pi d^2/4$ and therefore:

$$A = \pi d^2/4 = P_e/0.5 \, C_p \, \eta \, \rho \, U^3 \tag{2.22}$$

Rearranging, we obtain:

$$d = (8 \, P_e/\pi \, C_p \, \eta \, \rho \, U^3)^{0.5} \tag{2.23}$$

We may use this equation to estimate the rotor diameter required to produce a specified rated electrical power P_{er} at a stream velocity U_r – for example a machine designed to produce 1 MW at 2.5 m/s. We will assume the following typical values: $C_p = 0.5$; $\eta = 0.9$ and $\rho = 1030$, giving:

$$d = (8 \, P_{er}/1460 \, U_r^3)^{0.5} = 18.8 \text{ m} \tag{2.24}$$

This result is based on three simple assumptions: the amount of power in a tidal stream; the efficiency of capture by a turbine rotor; and the efficiency of conversion from mechanical to electrical power. It is interesting to note that the turbine is predicted to need a rotor diameter of 18.8 m, whereas a wind turbine rated at 1 MW has a typical diameter of about 45 m and nearly six times the swept area. This is consistent with Equation 2.23 bearing in mind that the density of air is about 800 times less than that of sea water; and that the rated speed of a wind turbine is typically about 13 m s^{-1} rather than 2.5 m s^{-1}.

Equation 2.24 can be used to predict the rotor diameters of tidal stream turbines with various electrical power ratings at a stream speed of 2.5 m s^{-1}. For example:

P_{er} (MW)	0.2	0.5	1	2
d (m)	8.4	13	19	27

Of course we can hardly expect our calculations to apply precisely in all cases. Not all turbines are designed to reach their rated power at stream speeds of 2.5 m s^{-1}, nor do they all work with the same efficiency. A 1 MW machine placed in a vigorous tidal stream might reach its rated power at a stream speed of 3 m s^{-1}, in which case its rotor diameter could probably be reduced to about 14 m – and so on. But overall, the analysis gives a fair indication of the rotor diameters to be expected in today's tidal stream turbines.

This leads us to a general point: one of the most important indications of a device's power generating capacity is its swept area. This applies to a wide range of turbine designs, VA as well as HA, and to other devices providing they have similar efficiencies. In other words, if you wish to make a quick estimate of a machine's rated power, first ascertain its swept area. The wind energy industry has produced many examples over the years of exaggerated claims about new designs (in extreme cases the advertised electrical output has exceeded the amount of power in the wind!). Fortunately the installers of large tidal stream devices are generally well informed and unlikely to be taken in by false promises. But caution is certainly in order and swept area will always be a key indicator of power generating capacity.

Some turbine designs employ tapering ducts to increase the effective swept area, accelerating the flow onto the rotor blades. It is sometimes claimed that a duct can increase efficiency dramatically because of the cubic relationship between stream power and velocity. But there is a fallacy here, because Bernoulli's equation tells us that speeding up a flow with a constriction does not alter its total energy content. The supposed benefit from accelerating the flow is cancelled out by a reduced pressure in the throat of the duct where the rotor is placed. Efficiency calculations should always be based on the total free-stream area 'tapped' by the device, which in this case includes the duct's cross-section. True, a duct allows the use of a smaller, higher-speed, rotor to yield a given power output; but against this must be set the capital cost of the duct – especially for a turbine in the megawatt class. A ducted turbine cannot be expected to outperform an open-rotor turbine intercepting the same total cross-section.

So far we have focussed on the power generated by a turbine at a particular stream velocity. But how does the power vary as the tidal stream flows backward and forward in response to the tides, and how can we assess the electrical energy produced over a tidal cycle? For a start we will re-write Equation 2.21 to allow for time-varying power and velocity:

$$P_e(t) = 0.5 \, C_p \, \eta \, \rho \, A \, U^3(t) \tag{2.25}$$

Figure 2.30 illustrates a typical set of curves for one tidal cycle in a location characterised by a semi-diurnal tidal pattern, for a turbine with power rating P_{er} at stream velocity U_r. The stream velocity $U(t)$ starts at zero and rises to a peak during the flood phase (between about 0 and 6 h). It reverses during the ebb phase (between about 6 and 12 h), but is here shown positive because we assume the turbine generates equally efficiently in both directions. Note that the peak flow velocity is slightly smaller on the ebb, a fairly common occurrence (in any case peak flow rates on successive tides are constantly changing in sympathy with the astronomical forces of the moon and sun; a turbine is hardly ever presented with identical flow patterns on successive

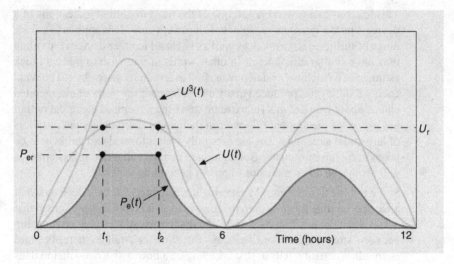

Figure 2.30 Tidal currents and electricity generation over a tidal cycle.

tides). In this example the flow rate exceeds the turbine's rated speed U_r between times t_1 and t_2 during the flood phase, but not during the ebb.

In many cases $U(t)$ has a roughly sinusoidal shape over each tidal half-cycle, and $U^3(t)$ may be approximated by a $\sin^3$ function. The turbine's electrical output $P_e(t)$, shown by the red curve, is proportional to $U^3(t)$ except that it limits automatically at its rated value P_{er} between times t_1 and t_2. Finally, the total electrical energy generated over the tidal cycle is given by the area under the power curve, shaded red in the figure. Adding together all such areas, for all tidal cycles throughout a year, gives the annual energy production. The ratio between this figure and the amount of energy that would be produced if the turbine operated continuously at its rated power is known as the *capacity factor*, the key performance measure already discussed in Section 1.4.

Figure 2.30 clarifies a number of important issues for turbine design and installation, especially the relationship between the rated output of a machine and the particular tidal stream in which it is to be deployed. The choice of rated velocity U_r has a big effect on the turbine's ability to take advantage of tidal half-cycles with relatively weak peak flows, compared with strong ones that exceed its power rating. The cubic relationship between stream velocity and power ensures that the most productive power peaks are almost certain to be partly 'wasted'. Over a complete year, especially in locations with complicated stream patterns showing pronounced daily, monthly and annual variations, all this requires careful planning and prediction of tidal stream patterns prior to installation.

2.4.2.2 Extracting energy: the Betz Limit

The rotor of a HA turbine intercepts the tidal stream and aims to transform as much of its kinetic energy as possible into useful mechanical energy. Although it is tempting to imagine that an ideal rotor could do this perfectly, a few moments' thought reveals a serious problem. To extract all the kinetic energy from a moving stream it must be brought to a complete standstill, yet this is impossible because the water has to pass through the rotor's swept area and continue its journey on the far side. It cannot 'pile up' in front of the rotor. So how much kinetic energy can actually be extracted? This fundamental question was tackled by Albert Betz (1885–1968), a German naval engineer and professor at the University of Göttingen who is best remembered for the theoretical limit to rotor efficiency he discovered.

The *Betz Limit* states that a rotor in a steady fluid stream – air or water – cannot convert more than 16/27, or 59%, of the stream's kinetic energy into mechanical energy. In other words its power coefficient cannot exceed 0.59. Rather intriguingly, this figure is based on the fundamentals of fluid dynamics and is independent of the precise type or design of the rotor. Betz' insight produced one of those moments when the elegance of theoretical physics meets practical engineering, issuing a warning to designers not to attempt the impossible. We need hardly be surprised that the practical difficulties of approaching a theoretical limit mean that large modern rotors achieve power coefficients up to about 0.5 or 50%, a figure we have used previously.

How was the Betz Limit derived? We start by considering the *stream tube* shown in Figure 2.31. Water enters the tube with a speed U_1 and, after passing through the rotor, exits with a lower speed U_2. The only flow is across the ends of the tube, and the speed reduction means that the tube's cross-sectional area must increase towards the downstream side. This is accompanied by a change in pressure across the rotor disc. There are some important assumptions:

- Water is incompressible and the flow homogeneous.
- There is no energy loss due to friction.
- The change in water pressure is uniform over the rotor disc (equivalent to specifying a rotor with an infinite number of blades).

A full analysis based on classical linear momentum theory can be used to calculate the rotor's efficiency, thrust and power output [16]. From the

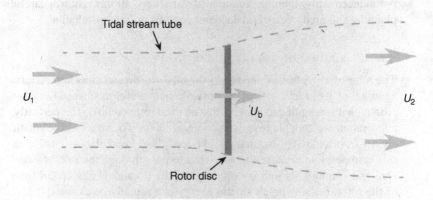

Figure 2.31 A rotor and its stream tube.

point of view of efficiency – our primary interest in this section – we may condense the argument considerably. First, it is clear that the rotor causes the reduction of stream velocity from U_1 to U_2 shown in Figure 2.31, and this occurs at the rotor plane. The average of these two velocities is therefore a good estimate of the velocity U_b actually 'seen' by the rotor blades:

$$U_b = (U_1 + U_2)/2 \tag{2.26}$$

We now introduce an *axial induction factor* (a), defined as the fractional decrease in velocity between the input to the stream tube and the rotor plane:

$$a = (U_1 - U_b)/U_1 = (U_1 - U_2)/2\,U_1 \tag{2.27}$$

giving:

$$U_b = (1-a)\,U_1 \tag{2.28}$$

If $a = 0$ the water passes through the rotor unhindered and no energy is captured. A value $a = 0.5$ implies that the velocity at the exit from the stream tube is zero, which is not allowed. So the Betz theory only holds over the range $0 \leq a < 0.5$.

The rotor's power coefficient C_p, equal to the fraction of the incoming power extracted, is related to the axial induction factor as follows [16]:

$$C_p = 4a\,(1 - a)^2 \tag{2.29}$$

What value of axial induction factor produces the greatest value of power coefficient? Differentiating Equation 2.29 and equating to zero, we obtain:

$$(3a - 1)(a - 1) = 0, \text{ giving } a = 1/3 \; or \; a = 1. \tag{2.30}$$

The first of these values represents the required maximum. Substituting $a = 1/3$ in Equation 2.29 we obtain Betz' famous result for the maximum value of power coefficient:

$$C_{\text{pmax}} = 4/3 \, (2/3)^2 = 16/27 = 59\% \tag{2.31}$$

An interesting aspect is the change in stream tube cross-section from the upstream to downstream side of the rotor. If $a = 1/3$, maximising the rotor power, the initial velocity U_1 is reduced by 1/3 at the rotor and by 2/3 at the exit from the stream tube. Since water is assumed incompressible any velocity changes must be offset by changes in cross-section. Therefore the stream tube must have an input cross-section equal to 2/3 of the rotor disc (swept area), expanding to twice the disc area downstream. It is only under these conditions, according to Betz, that the power coefficient can attain its maximum value.

You may feel that these results, widely quoted and respected, rest on slightly shaky foundations. It is hard for non-experts to assess the validity of the fluid dynamics involved, but we know that assumptions such as homogeneous flow through the rotor and absence of friction cannot truly apply in practice. This implies that the 59% is rather optimistic, but it has been accepted as the absolute upper bound on the efficiency of turbine rotors for about a century, and it seems clear that nobody ever expects to exceed it!

There is one more significant caveat, illustrated in Figure 2.32. This concerns the nature of the disturbed stream, or *wake*, downstream of the

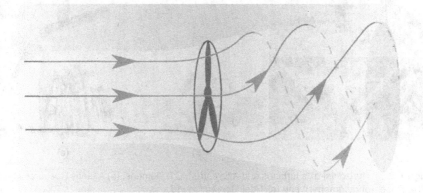

Figure 2.32 Vortex formation in a turbine wake.

rotor. Figure 2.31 has shown water leaving the stream tube parallel to the axis, at reduced speed but without any rotation. But as water passes through a tidal turbine and exerts torque on the rotor, there must be an opposite reaction on the stream. In other words there is some *vortex formation* in the wake, which imparts unwanted rotational kinetic energy to the stream and reduces the rotor's power coefficient. The greater the torque on the rotor, the stronger the vortex, so machines running at low speed with high torque are most affected. Fortunately vortex formation rarely subtracts more than a small percentage from the power coefficient of a well-designed rotor.

2.4.2.3 Lift and drag

The Betz Limit discussed in the previous section estimates the maximum efficiency of an ideal rotor in a steady stream as 59%. This famous result pays no attention to the actual design of the rotor, leaving such small details to the designers of blades and hubs! Our own story must pick up at this point since we wish to clarify the principles underlying today's tidal stream rotors [17], three examples of which are shown in Figure 2.33.

Two basic forces dictate the interaction between a tidal stream and the blades of a rotor, determining its hydrodynamic performance. The first is *lift*, a beneficial force on which all modern bladed turbines depend. And the second is *drag*, normally a detrimental force to be reduced as far as

| (a) | (b) | (c) |

Figure 2.33 Horizontal-axis turbines and rotors. (a) MCT/Siemens; (b) Atlantis Resources, Mike Roper (photographer); and (c) Tidal Generation Ltd.

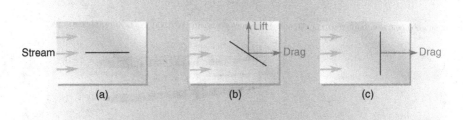

Figure 2.34 Forces on a thin plate in a steady stream.

possible – except in the case of machines such as the Savonius turbine (illustrated in Figure 2.29), which actually press it into service. A clear understanding of lift and drag is essential for appreciating the fundamental principles on which modern turbines depend.

We start by considering the forces acting on a thin, flat plate placed in a steady stream. Figure 2.34a shows an edge view, with the plate aligned exactly with the stream direction. Assuming the plate is very thin and smooth, there is negligible disturbance to the stream and no forces are generated.

In Figure 2.34b the plate is inclined at an angle to the oncoming stream, which tries to push it away with a drag force; and also to move it at right angles with a lift force. The net force on the plate is the resultant (vector sum) of these two forces. Figure 2.34c shows the plate positioned at right angles to the stream. There is no lift, but the drag force is large. We see that the relative amounts of lift and drag depend greatly on the inclination of the plate. To summarise:

- Lift forces are produced perpendicular to the oncoming stream.
- Drag forces are produced parallel to the oncoming stream.

Modern tidal stream turbines, like large wind turbines, are essentially lift machines. They use lift to generate forces at right angles to the stream, producing a twisting force or *torque* on the main shaft and delivering useful power. In effect they are entirely driven by lift forces; drag forces are a hindrance. Lift efficiently harnessed, and drag effectively minimised, are the key to success.

Figure 2.34b shows the flat plate inclined at an angle to the stream, generating both lift and drag. What it does not show is the turbulence and eddies produced on the plate's 'downstream' side, which accentuate drag. For a machine designed to operate on the lift principle with minimum drag,

Figure 2.35 Cambered wings with plenty of lift: the *Wright Flier* 2 takes to the sky in 1904 (Wikipedia).

special care must be taken to avoid turbulence by using a well-designed *hydrofoil section* instead of a simple flat plate.

The way in which lift forces act on rotor blades is subtle and somewhat mysterious to many people. Fortunately we can broach the topic gently by considering a more familiar situation – the forces acting on the wing of an aircraft .

All this has quite a history. By the time the Wright brothers made their powered flights at Kitty Hawk, North Carolina in the first decade of the last century, they were fully aware of the crucial importance of wing shape for maximising lift and minimising drag (Figure 2.35). Experiments carried out by a number of investigators since the late eighteenth century, and continued by the Wrights themselves, had established the efficacy of double-surfaced airfoils resembling the wings of a bird. Ever since the early days of powered flight aeronautical engineers have striven to perfect airfoil sections; and a parallel effort has been made by the designers of hydrofoil sections for water turbines.

Figure 2.36 shows a typical airfoil section through the wing of an aircraft facing a horizontal airstream. There are several important features:

- The airstream divides between the upper and lower surfaces. Since the upper surface is more curved (cambered) the air must travel further and faster over it, lowering the pressure. Conversely the lower airstream exerts positive pressure. Both effects contribute to lift.

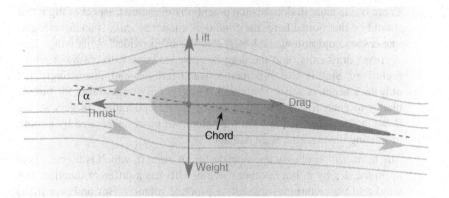

Figure 2.36 Forces acting on an aircraft wing in a horizontal airstream.

- The line between the centre of the airfoil's leading edge and the trailing edge is known as the *chord*. It is typically kept inclined to the airstream at a small angle (α) known as the *angle of attack*, to optimise lift.

- At low angles of attack the air flow over a well-designed smooth wing stays close to the wing surfaces, forming a *boundary layer*. However, if the angle of attack exceeds a certain value the boundary layer tends to separate, producing eddies and turbulence. At this point drag increases greatly, and lift reduces.

The four red arrows in the figure indicate the directions (but not the relative strengths) of four key forces. Assuming the aircraft is in level flight at constant speed:

- Lift (also produced by the other wing, and probably a tailplane) is at right angles to the oncoming airstream and exactly balances the weight of the aircraft.

- Drag (also on the other wing and the aircraft's fuselage) is exactly counteracted by the thrust of the engine(s).

If the lift increases, the aircraft gains height; if the drag increases, it loses speed. In level flight at constant speed, *lift equals weight and drag equals thrust*.

An important measure of an airfoil's quality is its *lift-to-drag ratio*. Ideally plenty of lift is provided to support the aircraft's weight but little engine power is needed to overcome drag, leading to good fuel economy. The greater the lift-to-drag ratio, the better.

There is one more important, and potentially disastrous, aspect of flight that should be mentioned here: the condition known as *stall*. If an aircraft gets into a 'nose-up' attitude at low speed, the lift provided by its wings may decrease drastically, and the drag increase, causing it to crash – a pilot's nightmare. Stall is also important to water turbines and, interestingly, not only in a negative sense. Although it is generally undesirable for turbine blades to enter a stall condition during normal operation, in very strong flows stall can be used to limit the amount of power generated by a rotor, protecting the machine from damage.

The term '*lift*' is clearly appropriate for an aircraft, which is literally held up in the sky by it, but in other machines lift has a different function. HA wind and water turbines use lift to produce rotation. But however lift is used, an efficient lift machine depends on a high lift-to-drag ratio.

We are now ready to explore the effects of lift and drag on a rotating turbine blade in some detail. Imagine a section through the hydrofoil, fairly near the hub, as shown in Figure 2.37. The stream flows horizontally from left to right and, viewed from the stream direction, the rotor is turning clockwise. Clearly, in this view the blade has a very different orientation from that of an aircraft wing in level flight, and this can be confusing. So let us start by emphasising that, viewed from the stream direction, the blade presents its

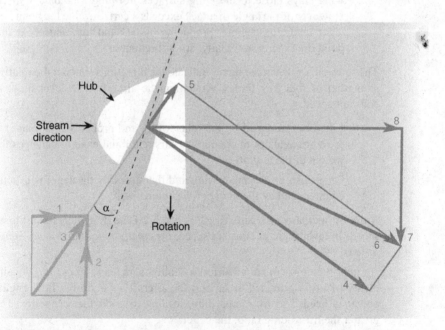

Figure 2.37 Flow speeds and forces on a rotating hydrofoil.

'underside' towards us – what, in an aircraft, would be its lower surface. The more curvy 'topside' is out of sight and it follows that the lift force is acting somewhat away from us and towards the back of the turbine.

There are other important distinctions between a water turbine and an aircraft in level flight. As the turbine rotates it *creates its own stream* at right angles to the 'natural' stream, and it is the *resultant stream* that acts upon the blades. Rather than experience the stream coming from the left side of the figure, the blades operate in a flow that comes more from the direction of rotation. This is illustrated in the figure by blue vectors 1, 2 and 3 that indicate stream strength as well as direction. They are:

1. The 'natural' stream at the turbine rotor, coming horizontally from the left.
2. The additional stream caused by rotation.
3. The resultant stream acting on the blade, equal to the vector sum of 1 and 2.

We now consider the forces generated, shown by numbered red and green vectors. The hydrofoil is inclined at a suitable angle of attack (α), generating plenty of lift (4) at right angles to the resultant stream. There is also some drag on the hydrofoil (5) and the net force it generates (6) is the vector sum of lift and drag (the amount of drag has been exaggerated here to show its effect clearly). And now comes a key point: we may resolve the net force (6) into two components; a desirable radial component (7) that produces blade rotation, and an unwelcome axial component (8) that tries to bend the blade towards the back of the turbine.

Figure 2.37 encapsulates and explains ideas that many people find challenging. In particular it shows that:

- To be effective the blade hydrofoil must be inclined at a suitable angle to the *resultant* stream.
- The direction of the resultant stream depends on the ratio between the speed of the natural stream and the speed of the turbine.
- Drag on the blade causes its net force (6) to move slightly away from the direction of the lift force (4) towards the back of the turbine, reducing the useful radial force (7) and increasing the undesirable axial force (8). This emphasises once again the value of a high lift-to-drag ratio.

A well-shaped hydrofoil with a rounded leading edge and sharp trailing edge can achieve a lift-to-drag ratio in excess of 50. Lift tends to increase linearly with angle of attack up to a certain angle – typically around 15° – above which separation of the boundary layer takes place and stall begins.

There is one further major distinction to be made between the operation of a turbine blade and an aircraft wing. So far we have considered a point on the blade fairly near the hub. But the stream speed caused by rotation equals Ωr, where Ω is the angular velocity of the rotor in radians per second and r is the radius considered (the distance from the centre of the hub). So, at a given speed of rotation, and with a constant incoming stream, the speed and direction of the resultant stream vary according to the distance from the hub. Near the blade tip, the resultant stream is largely due to the rotational component. This is quite different from an aircraft wing in level flight, which experiences the same wind speed and direction along its length. And the upshot is that, in order to optimise lift and maintain a suitable angle of attack along the length of a blade, it must be progressively *twisted* from hub to tip. In this respect a turbine rotor is very similar to a marine propeller.

Actually the situation is more complicated because tidal stream velocities are highly variable. The speed of the 'natural' stream impinging on a turbine rotor is always fluctuating in sympathy with the tides, altering the relationship between the 'natural' flow and the flow produced by rotation. In the next section we look at this effect in more detail and explain how it impacts on rotor speed and efficiency.

2.4.2.4 Rotor speed and power coefficient

How fast should a tidal stream turbine rotate? The answer to this basic question, which depends mainly on hydrodynamics, has important implications for the mechanical and electrical aspects of turbine design [17]. We have already seen that the Betz limit places a theoretical ceiling of 59% on the efficiency of a turbine rotor, regardless of its detailed design or speed of rotation. We have also discussed lift and drag forces, explaining how they act on a hydrofoil and the need for a blade to be progressively twisted from hub to tip. It is now time to relate these ideas to rotor speed and the way in which it may be varied to suit the tidal flow.

It is helpful to start by recalling the simple relationship between the power, speed and shaft torque of a rotating machine:

$$P = \Omega T = 2\pi N T \tag{2.32}$$

where P is the power measured in watts, Ω is the angular velocity in radians per second, T is the torque in newton metres (N m) and N is the speed in revolutions per second. For example, suppose a tidal stream turbine is generating 1 MW of mechanical power while turning once every 5 s (12 rpm). In this case P = 1×10^6 and N = 0.2, so the torque is :

$$T = P/2\pi N = 1 \times 10^6/0.4\pi = 0.8 \times 10^6 \text{ N m} \tag{2.33}$$

Figure 2.38 A large modern tidal stream turbine: high torque at low speed (Atlantis Resources).

We should pause for a moment to consider this enormous figure. A force of 1 N is close to 1/10th of a kilogram weight, so the twisting effort on the rotor shaft is about 80 000 kg m, or 80 t m. Imagine an Indian elephant weighing 1 t standing on the end of a long (and light) horizontal lever; the lever would need to be 80 m long for the animal to produce this amount of torque!

This example emphasises that large tidal stream turbines are quintessentially high-torque, low-speed machines, unusual in today's technological world (Figure 2.38). By comparison generators in conventional power plants usually run at 3000 rpm (or 3600 rpm in the USA). A machine running at 3000 rpm needs 250 times less torque than one running at 12 rpm, for the same amount of power. Ask any electrical or mechanical designer and he or she will confirm a general preference for high speed and low torque. Large low-speed turbines demand heavy shafts, expensive bearings and high-ratio gearboxes – or specially designed low-speed electrical generators.

Since power is proportional to the product of torque and speed, why not reduce the torque and increase the speed of a megawatt rotor? The answer lies mainly in the hydrodynamics of turbine blades, and we have already

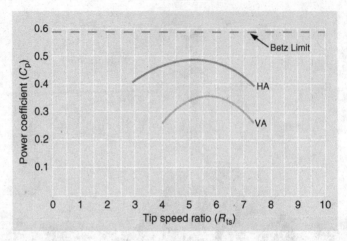

Figure 2.39 Variation of power coefficient with tip speed ratio.

hinted at it in the previous section: the ratio between the rotational speed of the blade tips and the speed of the incoming stream – the *tip speed ratio* (R_{ts}) – has a major influence on rotor efficiency. This is illustrated by Figure 2.39 which shows typical variations of power coefficient with tip speed ratio for large two-bladed and three-bladed HA turbines, and for VA turbines of the Darrieus type. The former achieve efficiencies close to 50% with tip speed ratios between about 4 and 6. The horizontal line at the top shows the theoretical maximum of 59% imposed by the Betz limit. It is difficult to explain the precise form of the HA curve, although we might note that if R_{ts} is too small a lot of water passes through the rotor without doing useful work on the blades; whereas if it is too large, each advancing blade encounters interference from the preceding one.

A turbine rotor with diameter d rotating at speed N has a tip speed of πdN. The tip speed ratio in a stream velocity U m s^{-1} is therefore:

$$R_{ts} = \pi dN/U \quad \text{so that} \quad N = R_{ts}\, U/\pi d \tag{2.34}$$

For example, a 1 MW turbine with a rotor diameter of 20 m operating at a tip speed ratio of 5 in a 2.5 m s^{-1} tidal stream rotates at:

$$N = 5 \times 2.5/20\,\pi = 0.20 \text{ rev s}^{-1}, \quad \text{equal to } 12 \text{ rpm} \tag{2.35}$$

It is worth noting that the tip speed in this example is 12.5 m s^{-1}. If it were to exceed about 15 m s^{-1} near the sea surface there would be danger of *cavitation* – the tendency for pockets of water vapour to form on the 'suction' side of a blade, or at the tip, due to the pressure dropping below the vapour pressure of water. Cavitation reduces efficiency and may damage the blade. It is fortunate that the tip speed ratio and cavitation limitations impose similar constraints on the maximum speed of the rotor.

In practice, tidal stream velocities change continuously during each tidal cycle. A turbine might be designed to start generating when the stream reaches 0.5 m s^{-1} and reach its rated output at 2.5 m s^{-1}. Ideally the rotor speed should track the stream velocity, keeping the tip speed ratio close to the optimum value and maximising energy capture. Conversely it is clear that a turbine with fixed blades, rotating at a constant speed, could only approach 50% rotor efficiency over a fairly narrow range of stream velocities.

HA rotors can either have *fixed* blades with the angle of attack and twist chosen as a compromise to suit the most common operating conditions. Or they can use *pitched* blades which swivel about their longitudinal axes to allow continuous optimisation of the angle of attack – an approach that is widely used in today's large wind turbines. The combination of variable rotor speed and pitched blades gives great operational flexibility, including [17]:

- Optimisation of the power coefficient over a wide range of stream velocities between the cut-in and rated values.
- Shedding excess power by 'blade pitching' whenever the stream velocity exceeds the turbine's rated value, reducing forces on the rotor and turbine structure.
- The ability to stop the turbine, even in a strong flow, by pitching the blades into a neutral or 'feathered' position.

The decision whether to go for fixed or pitched blades has many consequences [17]. With fixed blades, the complete turbine has either to be turned around to face the stream when the tidal flow reverses, or the blades must work well with the flow coming from either direction. If the latter, then the leading edge of a blade becomes the trailing edge when the flow reverses, implying that the shape of the hydrofoil must be symmetric – not the best recipe for a high lift-to-drag ratio. At the other extreme, fully-pitched blades that can swivel through 180° allow the blades to operate equally efficiently in either direction, with the added advantages mentioned above; but, of course, they are more expensive. There is presently no consensus over the best approach to adopt, balancing technical sophistication against cost, but tidal turbine technology may follow the wind industry in the coming decade and increasingly adopt pitched blades [16].

2.4.3 Turbine siting

It might seem fairly straightforward to lower a turbine into a tidal stream, fix it to a suitable foundation or mooring, connect a cable to shore, carry out routine maintenance, and do the occasional repair – until we remember the special features of tidal streams, especially those vigorous enough to promise a high electricity yield. An excellent overview of the problems and solutions is given elsewhere [17] and we will draw on it for a brief summary of some key issues.

We already discussed suitable sites for tidal stream turbines in Section 1.2.2 and, as examples, indicated some of Scotland's best resources in Figure 1.11. Needless to say, the most attractive ones from a power generation point of view have high peak and average flow velocities. But fast tidal races are tough environments for any form of machinery and would normally be avoided at all costs. A large turbine placed in a 4 m s^{-1} spring tidal current is subjected to drag forces equivalent to placing a similar device in winds of around 400 km h^{-1} – assuming such mega-hurricanes could be found anywhere on the planet. As the turbine generates electricity, its blades develop very large torques which, together with drag forces on the whole structure, produce large reaction forces on the moorings and seabed. We are talking here of forces upwards of 100 t for a 1 MW machine. Steady streams are hard enough to bear, but dynamic forces due to rotation and turbulence may be even tougher, leading to unwanted resonance and hastening fatigue failure. The fundamentals of resonance and fatigue are very similar for wave energy devices, and have been introduced in Section 2.2.3.2.

A strong tidal stream tends to scour the seabed, removing any sand, mud or loose material to expose a rocky surface which may be very uneven. Unfortunately, most rocks have limited load-bearing capability and may be subject to fracture, so proper anchoring of a turbine is crucial. There are three basic approaches:

1. Attachment to a rigid structure piled into the seabed. One possibility is the *monopile*, the type of single cylindrical steel tube hammered into the seabed that is extensively used to support offshore wind turbines. However, in practice, the support structure of a tidal turbine is generally fixed with several piles of smaller diameter.

2. Use of a gravity foundation which relies upon its weight to settle and remain secure on the seabed. Unfortunately the weight required is enormous – typically around 1000 t for a 1 MW machine, and scouring around the structure may undermine its stability.

3. Mooring a floating device using seabed anchors. But even here, all forces on the turbine structure must be transferred to, and resisted by, the seabed.

Closely related to the problem of anchorage is that of access and maintenance. How is the turbine to be brought above the surface for maintenance?

Will parts of the support structure be continuously 'surface breaking'? If so, what are the implications for navigation and public acceptability; if not, how will the turbine be accessed, especially bearing in mind that any underwater work must be restricted to brief periods of slack tide. These technical and operational issues have serious economic implications and it is hardly surprising that many solutions are currently being proposed. We shall meet some of them in Chapter 5.

Another important consideration is the height of a working turbine above the seabed. We discussed the von Kármán equation in Section 2.3.1, showing how the velocity of a tidal stream reduces as the seabed is approached. Even small reductions have a substantial effect on turbine output because of the cubic relationship between stream velocity and power; roughly speaking, 75% of the available energy is contained in the top half of the stream. From this point of view the turbine rotor should be well above the seabed. But this brings additional problems, including the need for taller support structures and the risk of interfering with shipping. Compromises are often needed. For example, a 1 MW turbine with a rotor diameter of about 20 m might be placed so that the highest point reached by the blades is 15 m below the sea surface at low water, with a clearance over the sea bed of 5 m. This implies a minimum water depth of 40 m. Generally the relationship between turbine size, height above the seabed, and water depth needs careful consideration – and impacts greatly on the design of the support structure.

So far we have considered individual turbines, but of course the eventual aim is to have large numbers grouped together in arrays. As in a wind farm, the basic task of each rotor is to extract as much kinetic energy from the moving stream as possible. In doing so, it creates wake and turbulence that may degrade the performance of neighbouring devices. Repeated over a large array, such effects give rise to *array losses* that reduce annual energy production [16]. So the question arises: how far should turbines be spaced apart? Clearly, generous spacing helps minimise array losses; but it also restricts the number of turbines that can be installed in a given area and increases cabling costs.

Figure 2.40 shows an array of turbines in relation to the stream direction. For simplicity they are shown arranged in a rectangular pattern. There are two spacings to be considered: *downstream* and *cross-stream*. Generous downstream spacing minimises interference by a turbine's wake on its downstream neighbours and reduces unwelcome stresses due to turbulence. Although there are no hard and fast rules, to judge from experience with wind turbines a value between 8 and 10 rotor diameters is generally sufficient to keep total array losses below 10%. For example, an array of 1 MW turbines with rotor diameters of 20 m might have a downstream spacing of about 200 m.

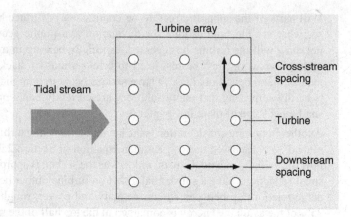

Figure 2.40 A turbine array.

Cross-stream spacing can be far smaller. Spacing by just over one rotor diameter might seem enough to avoid expensive arguments between adjacent devices, although the situation may be complicated by the generation of turbulence close to blade tips, and stream directions during ebb and flow phases may not be exactly 180° apart.

The power curve of a complete array – its power production as a function of the free stream velocity – differs from that of an individual turbine because array losses and turbulence affect turbines according to their position in the array. Power generation is generally slightly lower in downstream machines, and the stream velocity must exceed the rated value by some margin before all turbines reach their maximum power. The stream velocity may also vary considerably from one part of the array to another because of seabed bathymetry. Furthermore, turbulence causes the power output from turbines in different positions to fluctuate differently, and some turbines may be temporarily out of service. All these factors modify an array's power production, which is generally lower than from multiple isolated turbines. Of course, the layout of an array cannot generally follow the simple regular pattern of Figure 2.40, and careful on-site measurements are needed to establish optimal turbine placing, taking into account the bathymetry, the composition of the sea bed, and the flow characteristics of the tidal stream.

2.5 Research and development

2.5.1 Models and test tanks

For marine energy to make a significant contribution to national and global electricity supplies, wave and tidal stream devices have to reach a certain size and be installed in sufficient numbers. As with modern wind turbines, the power ratings of individual machines must reach megawatt scale; and if we take 1 GW of generation as a valuable addition to a nation's electricity supplies, and assume it is achieved using machines rated at 1 MW with an average capacity factor of 30%, over 3000 must be deployed. This gives some idea of the challenge faced by today's marine energy industry.

A machine capable of generating 1 MW of electricity from wave or tide does not come cheap. Millions of pounds or dollars are the order of the day. Designing a new machine from scratch relies on a great deal of brain power, computer simulation and experiments with scale models, before committing to heavy engineering and fabrication in steel and other materials. The process takes years. Good examples of painstaking and lengthy R&D programmes leading to successful full-scale prototypes are provided by the WECs and tidal stream turbines featured in Chapters 4 and 5.

A classic case of 'brain before brawn' in the early development of a new device is Professor Stephen Salter's 1970s research which led to his 'Nodding Duck' WEC (see Figure 1.14). Thirty five years later, in an account [10] that makes compulsive reading for anyone interested in the history of marine energy, he described how small balsa-wood models, self-designed instrumentation, and a simple water tank led to fundamental insights into the nature of waves and the ways in which their energy could be tapped. Today initial ideas are invariably subjected to numerical modelling [4] before committing substantial funds to physical devices. Research facilities, including computing power and instrumentation, have progressed enormously; yet there is still no substitute for the professional engineer's creative mind.

The design of effective water tanks for testing scale models is crucial to the R&D effort. Known as *wave tanks* when used for WECs, they use paddles and/or pistons to generate waves of various types [4]. Typically, flap paddles hinged at or near the tank floor produce waves of the deep-water variety in which the orbital motion of water particles decays exponentially with depth, appropriate for models of WECs that float in deep water. Pistons that move forward and backward can simulate waves in shallow water, relevant to testing models of near-shore devices mounted on the sea bed. A major wave tank requirement is to reduce unwanted reflections from the sides which interfere with the desired wave pattern, and much effort has been put into the design of 'artificial beaches' to absorb waves as they reach the far end. The number, size and control of paddles and pistons, and the physical dimensions of a tank, are extremely important considerations which determine the cost of the facility and the type of experiment it can handle. Testing is also very important for scale models of tidal turbines, but

Figure 2.41 Research and development of an innovative floating tidal stream turbine. (a,b) Tank testing a 1/20th scale rotor and a 1/16th scale device. (c, d) Moving to sea at 1/5th scale (Scotrenewables Tidal Power Ltd).

here the main requirement is a steady stream, normally achieved by pulling the model through stationary water in a long *towing tank*. Clearly, the tank size determines the scale of models that can be tested; on the whole, the bigger the better, except that most large wave and towing tanks are owned by universities and research institutes and their hire can be very expensive for device developers.

Tank testing allows a developer to move a design towards full-scale in a series of steps, honing it at each stage as experimental data is collected. A scale model saves weight as well as money, making it far easier and safer to transport and manoeuvre. Other things being equal a model at, say, 1/10th scale weighs only 1/1000th that of a full-scale prototype. As a design develops, larger models are often tested in sea conditions. The four photos in Figure 2.41 illustrate stages in the development of an innovative floating tidal stream device which is the subject of a case study in Chapter 5.

We have already explained in Section 2.3.1 that model results from tank tests may not, in general, be transferred directly to a full-scale prototype because the conditions are not *dynamically similar*. The problem arises whenever reduced-scale physical models are used to investigate new designs – in wind tunnel tests on aircraft and wind turbines as well as water tank tests on marine devices. It is resolved using two very important dimensionless quantities which relate dynamic conditions at reduced scale to those at full scale: the *Reynolds number* (already described in Section 2.3.1); and the *Froude number*. The Reynolds number is important for modelling tidal stream turbines, whereas the Froude number is more relevant to partially-submerged WECs that interact with surface waves.

William Froude (1810–1879), an English mathematician and engineer, pioneered the use of towing tanks to investigate the forces on a ship's hull as it moves through the water, with the aim of optimising hull design. He realised that forces on a model would only relate to those on the full-size ship if conditions in the two cases were dynamically similar, a criterion which Reynolds subsequently developed for submerged devices. We may summarise the distinction between Froude and Reynolds numbers by noting that there are three basic types of force acting on a fluid in motion – *inertial*, due to its mass or density; *viscous*, due to its viscosity or 'stickiness'; and *gravitational*. The ratio between inertial and viscous forces, which are dominant in a tidal stream, is represented in the Reynolds number. But if waves are involved the main forces are inertial and gravitational, and their ratio gives the Froude number. In a nutshell, when tank-testing a tidal stream model and relating results to full-scale, equalise Reynolds numbers; when testing WECs, equalise Froude numbers.

When design engineers are satisfied with the results of numerical and physical modelling a decision must be made to manufacture a full-scale prototype. Strong nerves are needed at this point because there are many pitfalls between model testing in tanks and deploying a megawatt device in real sea conditions. We devote the next section to the work of an organisation with an international reputation for progressing new wave and tidal stream devices from R&D towards successful commercialisation.

2.5.2   The European Marine Energy Centre (EMEC)

2.5.2.1   Wave and tidal test sites

The Orkney Islands, separated from the north-east tip of Scotland by the tidal races of the Pentland Firth and buffeted on their western coasts by Atlantic storms, are a wild and beautiful environment with huge potential for marine energy. There are some 70 islands altogether, covering nearly 1000 km^2, of which 17 are inhabited with a total population of about 20 000.

The two main towns, Kirkwall and Stromness, with populations of about 8000 and 2000, respectively, are on the island known as Mainland. Orkney has gained wide experience in the energy industries since an oil terminal was opened in the 1970s, including the testing of a giant wind turbine in the 1980s, and there are now many local companies and organisations exporting their knowledge and skills around the world. Orkney has the most advanced, and geographically concentrated, deployment of wave and tidal stream energy technology and associated infrastructure in the world, and in 2012 the UK government announced the establishment of a Pentland Firth and Orkney Waters marine energy park. It is estimated that the region could generate 18 TWh of renewable energy annually and this is reflected in the location of EMEC, widely acknowledged as the international leader in the testing of full-scale wave and tidal devices in ocean conditions.

EMEC, with headquarters in Stromness, offers developers two full-scale test sites. *Billia Croo*, the wave test area lying off the west coast of Mainland, receives the full force of swells arriving across the north Atlantic. The *Fall of Warness* tidal test area, a comparatively sheltered location some 20 km north of Kirkwall off the island of Eday, is famous for its vigorous

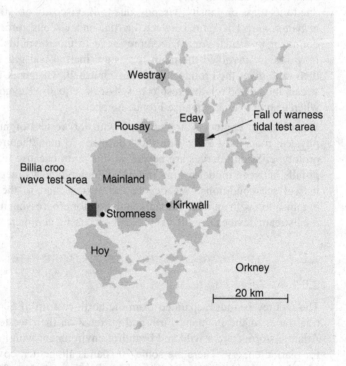

Figure 2.42 The Orkney Islands and full-scale EMEC test areas.

tidal streams (see Figure 2.42). Taken together, the two locations offer incomparable facilities to device developers.

Figure 2.43 shows an aerial view of the Billia Croo site – in rather gentler conditions than usual – and the site layout is shown in Figure 2.44. Its location and facilities offer developers a full range of ocean conditions, including uninterrupted Atlantic swells. Incoming waves, predominantly from a westerly or north-westerly direction, have an average significant height of about 1.9 m, an average period of about 6 s, and an average power over the year of about 22 kW m^{-1}. In extreme conditions the SWH may reach 15 m with an average period of about 14 s. Five tests berths about 2 km offshore in water depths of around 50 m are connected to EMEC's onshore control and switching station by custom-built seabed cables, buried as they approach the shoreline (see Figure 2.45). The cables incorporate fibre-optic bundles for data transmission. A smaller area close to shore is available for testing shallow-water devices.

Assessing the performance of devices depends upon accurate monitoring of wave conditions at the test berths. Two directional Waverider buoys (see Section 2.1.5) are deployed in the test area to provide continuously updated measurements of wave height, period, direction and power, transmitted by radio link to EMEC's data centre and recorded by a *SCADA* (*supervisory*

Figure 2.43 The Billia Croo wave test area (EMEC).

WAVE TEST SITE

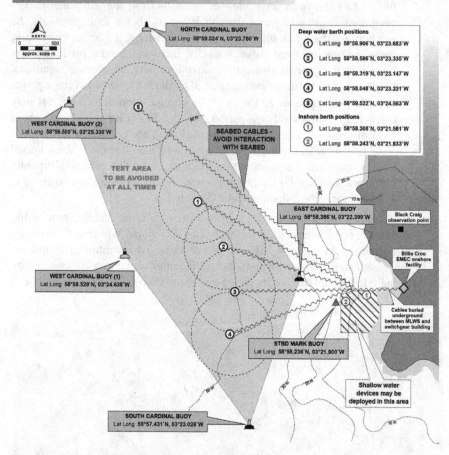

Figure 2.44 Layout of facilities at the Billia Croo site (EMEC).

control and data acquisition) system. A typical record of wave heights obtained from one of the buoys over a period of about two days is shown in Figure 2.46. In addition closed-circuit TV is installed at the Black Craig observation point high on the cliffs to the east of the site to allow visual monitoring, and weather data is obtained from an onshore met station.

EMEC's other full-scale facility is the tidal test site at Fall of Warness. Figure 2.47 shows an aerial view with the island of Eday in the distance. You may be able to detect the strong tidal stream in the narrow channel between Eday and the much smaller island of Muckle Green Holm in

Figure 2.45 The wave substation at Billia Croo (EMEC).

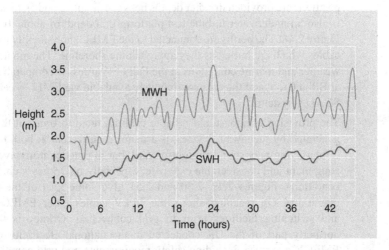

Figure 2.46 A record of significant wave height (SWH) and maximum wave height (MWH) at Billia Croo. (Redrawn from EMEC data.)

Figure 2.47 The tidal test area at Fall of Warness (EMEC).

the middle distance. Spring tides produce streams of about 3.5 m s^{-1} in a south-easterly direction and about 3.3 m s^{-1} in a north-westerly direction, reducing at neap tides to around 1.5 and 1.3 m s^{-1}, respectively. Eight berth positions are provided in water depths between about 35 and 50 m, and there is also an underwater turbine test platform in a depth of about 10 m – see Figure 2.48. The berths are connected to the EMEC shore facility by seabed cables which are buried as they approach the shoreline. The monitoring of weather and marine conditions at the Fall of Warness is comparable to that at Billia Croo, and the data continuously available via EMEC's centralised SCADA system.

The provision of robust and reliable cabling, rated to transmit the power generated by megawatt devices, is a major requirement at both test sites. Cable-laying may sound straightforward, but in practice it involves heavy equipment and considerable expertise, especially in Orkney's challenging conditions. Figures 2.49, 2.50 and 2.51 give some idea of the scale of operations. On reaching shore, the 11 kV cables enter EMEC control and switching facilities prior to grid connection. Orkney is the most northerly part of the UK connected to the national electricity grid and EMEC's expertise, including safety requirements and grid compliance, is a key advantage offered to developers. Grid connection also means that developers can earn a return for any electricity they generate.

TIDAL TEST SITE

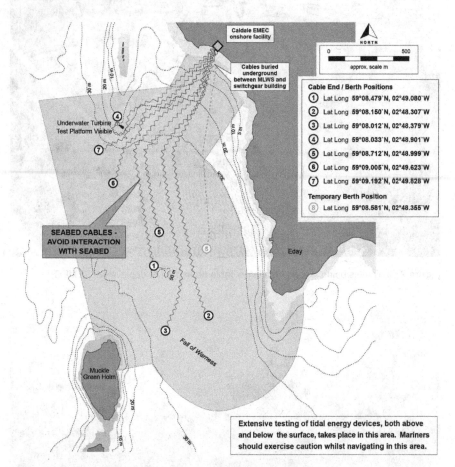

Figure 2.48 Layout of the tidal test area at Fall of Warness (EMEC).

In addition to its main test areas, EMEC has two 'scale sites' with facilities for testing smaller-scale devices and/or components in gentler conditions. The scale wave site is located about 10 km south of Kirkwall; the tidal scale site is a few kilometres north-east of Kirkwall in Shapinsay Sound. Both offer accessible real-sea conditions without needing the large support vessels and plant used to deploy full-scale devices at the two full-scale sites. A specially-designed test support buoy is moored at each location

117

Figure 2.49 Cables being loaded on a large cable drum prior to deployment (EMEC).

Figure 2.50 Cables come ashore at the tidal test site (EMEC).

Figure 2.51 Marine operations at the tidal test site (EMEC).

Figure 2.52 Test Support Buoy at the scale tidal site (EMEC).

Figure 2.53 Transport of ballast block moorings to the scale tidal site (EMEC).

(see Figure 2.52), offering electrical power to devices on test, dissipating any unwanted electricity generation, and transmitting wireless data to shore for performance monitoring. Ballast blocks are provided as moorings (see Figure 2.53) and the sites' energy resources are monitored by Waverider buoys.

During initial set-up of the wave and tidal test sites *Environmental Impact Assessments (EIAs)* were undertaken, characterising the sites and identifying any sensitive species which would need monitoring. Since then great care has been taken to monitor the marine species passing through the sites as more and more developers make use of the facilities, and the collected data are used to review and modify procedures where necessary. Prior to deployment, developers are required to consult over any likely impacts of their devices on the local marine environment and on navigation. EMEC guides clients through the consenting process and submits applications for the necessary licences. Many of the devices described in Chapters 4 and 5 are currently, or have been, undergoing tests at EMEC. The full-scale wave and tide test facilities are officially registered with the United Kingdom Accreditation Service (UKAS), enabling EMEC to provide internationally recognised verification of device performance.

2.5.2.2 Research activities

EMEC's main role is to provide operational testing and monitoring of marine energy devices, but it is also increasingly involved in research. Links with many developers, academic institutions, government departments and regulatory bodies put it in a unique position to pursue research on a number of fronts, making use of the exceptional wave and tidal sites and their activities. The sites are effectively large-scale laboratories where a large amount of data is collected from novel devices, many of which are being tested at full-scale and grid-connected for the first time. Comprehensive real-time measurement and recording of simultaneous meteorological and oceanic conditions – the so-called *metocean* climate – is extremely valuable to developers wishing to assess the detailed technical performance of their devices. The sites also offer scope for environmental investigations of great interest to the international marine energy community. Concerns about potential environmental effects of devices are a key area of concern for regulators in the early stages of this industry and this is reflected in the research emphasis to date in this area.

Broadly speaking the research programme may be divided into wave, tidal and more general topics, and the selection given here is designed to illustrate the broad range of work that has been or is being undertaken, most of which involves collaboration with project sponsors and partners [7].

Recent and current research interests and activities based at Billia Croo, EMEC's full-scale wave test area, include the following (see Figure 2.54):

- *Resource assessment.* Real-time wave and meteorological data are continuously collected and published in monthly reports for the benefit of developers assessing the performance of their devices. EMEC also cooperates with higher education institutions to investigate the details of wave climate at the site, including high-resolution numerical modelling.

- *Surface interactions with wave devices.* A high-resolution camera at the Black Craig observation point on a cliff overlooking the site is available for recording the frequency and nature of interactions between waves and the surface-piercing parts of wave devices. This information can be used in the design and development of devices, and for contributing to assessments of possible impacts on wildlife.

- *Underwater acoustic environment.* Underwater noise is important to cetaceans that rely on echo-location for feeding and communication. It may also disturb diving birds and some fish species. Any noise generated by wave devices adds to the background noise caused by breaking waves, ship engines and other marine wildlife.

121

Careful study of the relative strengths and characteristics of all these sources helps clarify the likelihood of devices having any impact on wildlife.

- *Wildlife observation and conservation.* Monitoring the wildlife visible at the sea surface at Billia Croo gives valuable data on the number of marine and bird species and individuals at the site and the ways in which they use it. This information helps developers and regulators assess the likelihood of possible interactions between energy devices and marine wildlife.

- *Investigation of lobster distribution at the wave site.* Marine energy devices could have beneficial effects on local stocks of fish and crustaceans since some of the device structures may provide suitable habitats for a variety of species. One research project involves rearing tens of thousands of juvenile lobsters in a hatchery, tagging and releasing them into the test area and subsequently monitoring their interactions with the local lobster population within a working test site. The project also sets the scene for further research in the future.

Research based at Fall of Warness, the full-scale tidal test area, includes the following:

- *Resource assessment and energy extraction.* Tidal streams are not homogeneous and are greatly affected by coastal geography and bathymetry. Although the site is fairly protected from oceanic conditions, waves can propagate into it from the northwest and southeast and very steep waves may be produced when tide and waves run against each other. Waves and wind tend to perturb tidal stream velocities, producing turbulence and complicating the conditions in which tidal devices operate. A wide range of such effects is being studied by computer modelling backed up by advanced measurement techniques, giving a comprehensive picture of the raw energy available to tidal devices.

- *Interactions with wildlife.* Acoustic noise from tidal devices may adversely affect certain species of sea mammals, fish and diving seabirds, so it is important to measure a site's background noise, establishing a baseline for comparison with device-generated noise. Obtaining accurate measurements can be challenging in areas of high tidal flow, since there is a lot of noise associated with water flowing past measuring equipment, which can artificially add to the noise being measured. Methods and equipment for gathering the data have been developed, and baseline characterisations produced. Damage due to physical collision is also possible, but may be hard to observe

Figure 2.54 Examples of research activities: observations above Billia Croo (a, b); high-tech ocean mapping (c, d) and lobster conservation (e, f) (EMEC).

due to water turbidity and insufficient natural light. Special sonar techniques are therefore employed to investigate possible collision damage. It is also important to discover whether the presence and/or operation of devices affect the distribution and behaviour of wildlife.

123

EMEC is also involved in a wide range of more general R&D activities including the development of streamlined consenting processes for marine renewables in Scotland, and numerical modelling of the complete waters around Orkney. Taken together, these projects confirm it as a world leader in marine energy research.

References

1. M.E. McCormick. *Ocean Wave Energy Conversion*, Dover Publications: New York (2007).
2. J. Falnes and J. Hals. *Wave Energy and its Utilisation: A Contribution to the EU Leonardo Pilot Project 1860*, NTNU: Trondheim (1999).
3. T. Ainsworth. *Significant Wave Height: A Closer Look at Wave Forecasts*, NOAA (NWS): Alaska (2006).
4. J. Cruz, ed. *Ocean Wave Energy: Current Status and Future Perspectives*, Springer: Berlin (2008).
5. Datawell BV, www.datawell.nl (accessed 24 April 2013).
6. P.A. Lynn and W. Fuerst. *Introductory Digital Signal Processing with Computer Applications*, John Wiley & Sons, Ltd: Chichester (1999).
7. EMEC, www.emec.org.uk (accessed 24 April 2013).
8. S. Salter. Wave power. *Nature*, **249**, 720–724 (1974).
9. Voith Hydro Wavegen Ltd, www.wavegen.co.uk (accessed 24 April 2013).
10. S. Salter. *Looking Back*, Chapter 2 in ref. 4 above, Springer: Berlin (2008).
11. J. Hardisty. *The Analysis of Tidal Stream Power*, Wiley-Blackwell: Chichester (2009).
12. A.T. Doodson. The harmonic development of the tide-generating potential, *Proceedings of the Royal Society*, **100** (704), 305–329 (1921).
13. R.H. Stewart. *Introduction to Physical Oceanography*, Texas A & M University (2005).
14. Tides and Currents, www.tidesandcurrents.noaa.gov (accessed 24 April 2013).
15. The United Kingdom Hydrographic Office, www.ukho.gov.uk (accessed 24 April 2013).
16. J.F. Manwell, J.G. McGowan, and A.L. Rogers. *Wind Energy Explained: Theory, Design and Application*, 2nd edn, John Wiley & Sons, Ltd: Chichester (2009).
17. Fraenkel, P. *Practical Tidal Turbine Design Considerations: A Review of Technical Alternatives and Key Design Decisions Leading to the Development of the SeaGen 1.2 MW Tidal Turbine*. Ocean Power Fluid Machinery Seminar, Institution of Mechanical Engineers, London (2010).

3 Generating electricity

3.1 Introductory

The development of wave and tidal-stream energy has reached the stage where individual devices, rated at about 1 MW, are being connected to electricity grids and, in the coming years, we may confidently expect arrays of devices to make significant contributions to regional, or even national, electricity supplies. As we explained in Section 1.4, a 1 MW machine operating with a typical capacity factor of 35% generates electricity over the course of a year equivalent to the needs of about 600 households in Western Europe. A 100 MW array might service 60 000 households or about 200 000 people. Yet few people live close to high waves and vigorous tidal streams. To take just one example, the population of the Orkney Islands is around 20 000; and if Orkney is to become a powerhouse of marine energy, its renewable electricity will have to find its way reliably and efficiently to mainland Scotland and the large electricity grid that serves it.

We have become so used to the idea of conventional power plants feeding electricity into grid networks that we take the process for granted. It is hardly surprising if designers of wave and tidal machines, many of whom are highly expert in mechanical engineering and hydrodynamics, tend to assume that any mechanical power they generate can simply 'be converted into electricity and fed into the grid'. Perhaps it can, but the process is not entirely simple, especially when generation is intermittent with occasional peaks that an existing electricity network may struggle to accommodate. Furthermore, the vast majority of modern electricity grids and networks use alternating current (AC) electricity, which has its own strange set of ideas.

Electricity from Wave and Tide: An Introduction to Marine Energy, First Edition. Paul A. Lynn.
© 2014 John Wiley & Sons, Ltd. Published 2014 by John Wiley & Sons, Ltd.

To many people AC electrical engineering is something of a black art full of obscure terminology and it can be a struggle to find a book or web site that explains it adequately to the non-specialist. So in this chapter we aim to shed light on such AC curiosities as reactance and impedance, real and reactive power and power factor correction; to introduce the operating principles of electrical generators and to explain how modern electronic control eases the problem of converting unruly marine energy into the well-behaved electricity required by an electricity grid. Finally, we introduce some key ideas about the operation and reliability of large grid networks. All these issues have been exercising minds in the wind energy industry for many years [1] and the approach adopted here is very similar to that in a book on wind energy [2] by the present author, but with copious minor alterations to reflect the shift of focus towards wave and tide.

However, we must be careful not to jump the gun. The first challenge is to capture the raw energy of wave or tide and convert it to a form suitable for driving an electrical generator. In other words a suitable system must be designed for efficient *power take-off*, a topic which we tackle in the following section.

3.2   Power take-off

The design of efficient power take-off systems is generally more challenging for wave energy converters (WECs) than tidal-stream devices. As discussed in Section 2.2.2 and illustrated in Figure 2.15, wave devices take many forms and require different approaches to power take-off. For example, attenuators and oscillating wave surge converters typically convert the raw energy of waves into hydraulic energy (high-pressure oil or water) which is then used to drive a hydraulic motor coupled to an electrical generator; oscillating water column (OWC) devices convert it into pneumatic energy (compressed air). By comparison the majority of tidal stream devices, like wind turbines, use a rotor and gearbox to drive an electrical generator directly.

The value of hydraulics or pneumatics as an intermediate stage for power take-off in a WEC is well described elsewhere [3]; here we settle for a much shorter account that brings out some of the main points. We start by noting that electrical engineers responsible for generating electricity in conventional power plants like nothing better than an energy source – fossil fuel, hydro or nuclear – that can deliver steady power to a turbine and electrical generator running at constant speed. The intermittent, variable and somewhat unpredictable power derived from renewable sources is very different. Wave energy, in particular, has a number of awkward features and we can hardly do better than repeat some remarks by Professor Stephen Salter which are also quoted in the above reference:

It is easy to make a device that will respond vigorously to the action of sea waves. Indeed it is quite hard to make one that will not. However the conversion of the slow, random, reversing energy flows with very high extreme values into phase-locked synchronous electricity with power quality acceptable to a utility network is very much harder.

The motion of most wave devices is oscillatory, in sympathy with the waves that power them, and the oscillations tend to be very slow. The large forces and low velocities involved, which are similar to those encountered in heavy lifting equipment, metal presses, and excavators, suggest high-pressure hydraulics as an ideal power take-off technology. Another major consideration is the 'lumpy' nature of wave energy, especially in vigorous random seas, with big power fluctuations from one wave to the next and occasional high peaks. To convert such an unruly source into a reasonably steady output requires a significant amount of short-term energy storage and this, too, may be conveniently provided by hydraulics in the form of high-pressure storage tanks or 'accumulators'. Finally, there is the control problem – the need to regulate random forces and motions to optimise energy capture while ensuring a device's survival. Hydraulics coupled to computers are well-suited to sophisticated control tasks, as suggested by their widespread use for controlling aircraft flight surfaces, ship steering and industrial robots.

Figure 3.1 illustrates a basic scheme. Wave motion causes backward and forward movements of a piston within a hydraulic cylinder (C), forcing oil into the high-pressure (HP) line of the circuit via a control valve (V). An accumulator (A) provides short-term energy storage by compression of

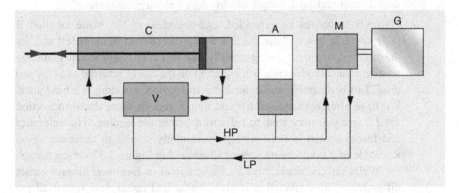

Figure 3.1 A basic hydraulic system for power take-off.

a gas, often nitrogen, and energy release by decompression, expelling oil back into the circuit. The accumulator's essential role is to isolate the highly variable energy of individual waves and produce a reasonably steady input to a hydraulic motor (M) and electrical generator (G). Oil is returned from the motor via the control valve to the cylinder at low pressure (LP). Note that the motor and generator are high-speed low-torque units compared with the slow, forceful oscillations of the piston. In effect, the hydraulic system performs a smoothing and 'gearing up' of the wave motion, equivalent to the use of a flywheel and high-ratio gearbox in a mechanical system. Many variations of this basic scheme have been proposed for WECs over the years and we will meet examples in Chapter 4.

A very different type of power take-off involves the use of air turbines in OWC devices, as already discussed in Section 2.2.2. A historical example of such a system was illustrated in Figure 1.13b. Incoming waves alternately pressurise and depressurise air in a chamber above the water column, causing air to flow backwards and forwards through the turbine rotor. A simple but effective type of turbine, developed in the 1970s by Professor Alan Wells at Queen's University Belfast, is widely recommended for this role due to its 'self-rectifying' properties. A *Wells turbine* speeds up and slows down during each air pressure cycle, but never reverses, making it suitable for running a conventional electrical generator.

The operating principle of the Wells turbine is illustrated in Figure 3.2. Figure 3.2a shows a rotor consisting of a number of airfoil blades mounted around a central disk. Air passes axially through the rotor in either direction, producing unidirectional rotation. For this to occur, the blade airfoil must be symmetrical in form and arranged so that the chord line joining its leading and trailing edges lies in the plane of rotation. However, it is not immediately obvious how a set of blades arranged in this way can generate useful, unidirectional, torque for driving an electrical generator.

Figure 3.2b shows an expanded, end-on, view of the blade labelled B in Figure 3.2a. We will assume that the air direction is from the left and, for the moment, that it is flowing steadily. We first need to appreciate that there are two basic aerodynamic forces acting on the blade, referred to as *lift* and *drag*. Lift is desirable and contributes to rotation, but drag is a hindrance. We have already discussed lift and drag forces in some detail in Section 2.4.2.3 and you may wish to refer to it before proceeding. The velocities and forces shown in the figure are essentially similar to those acting on the blade of a tidal stream turbine illustrated in Figure 2.37 – even though the Wells turbine blade must be symmetrical in form and aligned rather differently. The explanation is also similar and rather than repeat all the details here we will simply emphasise the main points.

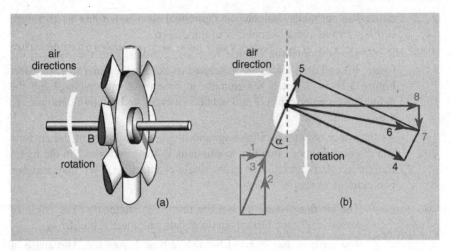

Figure 3.2 The Wells turbine.

As the Wells turbine rotates it *creates its own airstream* at right angles to the incoming air direction, and it is the *resultant airstream* that acts upon the blade. Rather than experience the air coming from the left side of the figure, the blade operates in a flow that comes more from the direction of rotation. This is illustrated in the figure by blue vectors 1, 2 and 3 that indicate airstream velocity and direction. They are:

1. The incoming stream at the turbine rotor, coming horizontally from the left.

2. The additional stream caused by rotation.

3. The resultant stream acting on the blade, equal to the vector sum of 1 and 2.

The various forces generated are shown by numbered red and green vectors. The blade is inclined at an angle of attack (α), producing lift (4) at right angles to the resultant stream. There is also some drag on the blade (5). The net force (6) is the vector sum of lift and drag (the amount of drag has been exaggerated here to show its effect clearly). And now comes a key point: we may resolve the net force (6) into two components: a desirable radial component (7) that produces blade rotation, and an unproductive axial component (8) that tries to bend the blade backwards. Note how drag causes the net force (6) to move slightly away from the direction of the lift force (4), reducing the useful radial force (7) and increasing the unwanted axial force (8). This emphasises the value of a high lift-to-drag ratio. Actually the symmetrical airfoils used in a Wells turbine have lower ratios than optimally designed asymmetrical airfoils, but this disadvantage is offset by the overall simplicity of the design.

If the air flow into the turbine reverses, coming from the right in the figure, lift and drag forces are generated in exactly the same way. In effect Figure 3.2b is 'flipped' horizontally to produce a mirror-image but the radial force component (7) still acts downwards and rotation continues in the same direction.

As the water column oscillates up and down, the velocity of the air flow and the turbine speed fluctuate continuously. Vectors 1 and 2 in the figure change in relative size, altering the angle of attack (α). We may imagine two extreme cases:

- The air flow continues but the turbine is stationary. The angle of attack is 90°, the blade is in deep stall, and there is no lift.
- The air flow has stopped but the turbine is rotating. The angle of attack is 0° and, because the airfoil is symmetrical, there is no lift.

In practice, the blades are most effective at angles of attack around 6–8°. Outside a range of about 2–15° the turbine cannot operate. As the water level in the OWC rises and falls in response to the waves, the speed of the turbine must be controlled in sympathy, ideally maintaining an optimum angle of attack and generating maximum power. Much theoretical work has been done over the years on Wells turbines, not least because a pioneering OWC on the Isle of Islay in Scotland was grid-connected in 2000 and has since provided a wealth of performance data on this type of device (see Section 4.2.3).

One other approach to power take-off has received special attention from the wave energy community – using wave motion to drive directly an unconventional linear (as opposed to rotary) electrical generator. We will reserve this topic until we have introduced AC electricity and electrical generators in the following sections.

3.3 AC electricity

Today's electricity grids are almost invariably based on AC electricity. The early years of the electrical power industry saw intense competition between AC systems championed by Nikola Tesla (1856–1943) (Figure 3.3a) and DC or *direct current*, systems preferred by Thomas Edison (1847–1931) (Figure 3.3b). However the race was eventually won by AC, which came to dominate the generation, transmission, distribution and utilisation of electrical energy during the twentieth century. Tesla, born in what is now

Figure 3.3 Pioneers of electrical power: (a) Tesla and (b) Edison (Wikipedia).

Croatia, spent most of his life in New York and is particularly remembered for his invention of the AC induction motor that today drives much of global industry. Its cousin, the induction generator, is widely specified for delivering renewable power to electricity grids.

One of the principal advantages of AC electricity is that its voltage level may be easily changed using transformers. Typically, AC is generated at one voltage level, transformed up for long-distance transmission and distribution, and finally transformed down to a safe level for use by consumers. Whenever the voltage is raised, currents are reduced and vice versa. Transmission at high voltage (and low current) reduces losses in power lines, allowing the use of much smaller and less expensive conductors. At the consumer end, AC is well suited to running cheap and efficient electric motors in a huge variety of applications, and even keeps our electric clocks running on time!

However, we must be careful not dismiss DC power transmission. The power-carrying capacity of electric cables is generally greater when fed with DC rather than AC and DC is sometimes favoured for long-distance transmission of bulk power from one region or country to another. DC also has special advantages for long submarine links, including cables to deliver large-scale offshore renewable electricity to the land.

In an AC system, voltages and currents in linear circuits and devices vary *sinusoidally*. Currents flow backwards and forwards, changing direction at a

131

rate determined by the grid frequency – for example 50 cycles per second or Hertz (Hz) in Europe and 60 Hz in North America. However, it is important to realise that the voltage and current waveforms, although sinusoidal in shape, are not generally *in phase* with each other, and this has profound implications for the generation and use of grid electricity. Nonlinear circuits, which include electronic switches used for controlling wind and tidal turbines, introduce non-sinusoidal waveforms and their analysis is more complicated. For the time being we will focus on linear circuits because they illustrate and explain most key aspects of AC electricity.

In general, the time-varying waveform $v(t)$ of an AC voltage may be expressed as:

$$v(t) = V_p \sin(\omega t + \varphi) = V_p \sin(2\pi f t + \varphi) \tag{3.1}$$

where V_p is the peak value (amplitude), ω is the frequency in radians per second, f is the frequency in Hz, t is the time in seconds and φ is the phase angle. If $\varphi = 0$ the function is a sine wave that passes through zero at $t = 0$. This may be thought of as a 'reference' waveform. Other waveforms of voltage and current with different phase angles are shifted along the time axis with respect to the reference.

We now consider voltage and current waveforms for three basic types of linear circuit element: *resistors, inductors and capacitors*. Although you may not often meet them as individual elements, the ideas and terminology involved are crucial for understanding the generation and distribution of AC electrical power by marine energy devices. Figure 3.4a shows two cycles, or *periods*, of a continuous voltage waveform – a sine function second with peak value V_p and period T (equal to the reciprocal of the frequency in Hz). If this AC voltage is applied to a resistor of value $R\ \Omega$, as in Figure 3.4b, the current that flows is proportional to the voltage at every instant and there is no phase shift. By Ohm's law its peak value is V_p/R. The instantaneous power dissipated in the resistor equals the product of voltage and current:

$$p_r(t) = V_p \sin(\omega t)\ V_p/R\ \sin(\omega t) = (V_p^2/2R)\ (1 - \cos(2\omega t)) \tag{3.2}$$

We see that the power fluctuates cosinusoidally, but at twice the frequency of the voltage source and with an offset that keeps it positive. The area under the power curve, shaded red in the figure, represents the energy supplied by the AC source and dissipated in the resistor. In some ways this is all rather surprising. After all, if we turn on an electric heater or an incandescent lamp – essentially both resistors – it is not obvious that the power is fluctuating at twice the supply frequency; but this is because any resulting 'flicker' is too fast for our eyes and brains to follow.

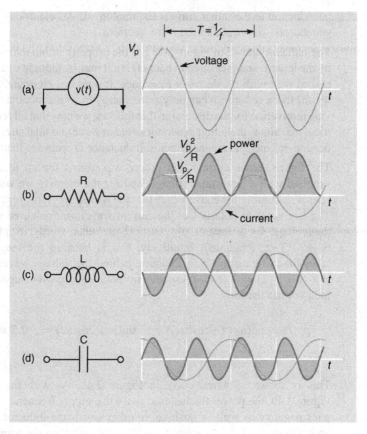

Figure 3.4 AC voltage, current and power.

It is clear from the figure that the *average* power dissipated in the resistor is half the peak power, that is $V_p^2/2R$. The same amount of power would be dissipated if it were connected to a DC voltage of $V_p/\sqrt{2} = 0.7071V_p$. In other words the heating effect of a sinusoidal voltage with peak value V_p is the same as that of a DC voltage of $V_p/\sqrt{2}$. Since heating is, in an important sense, what matters (and what we pay for when we turn on a heater!) the voltage of an AC supply is generally quoted as $V_p/\sqrt{2}$, referred to as the *rms* (*root mean square*) value. For example, the '220 V' of a European domestic supply refers to its rms value; the peak value is $220\sqrt{2} = 311$ V.

AC electricity becomes rather more interesting when we consider inductors and capacitors because they produce phase shifts between voltage and current waveforms. Understanding these phase shifts forms an essential introduction to the rather curious terminology of AC electricity and power generation – as we shall see in later sections.

The *inductor* shown in Figure 3.4c has the property of *inductance*, denoted by the letter L and measured in henrys (H). It may be thought of as an 'ideal' coil of wire with no electrical resistance. It does not *dissipate* energy (get warm) like a resistor; it can only *store* energy in its magnetic field. To provide motivation for the discussion that follows, we note that all rotating electrical machines, including generators used in wave and tidal stream devices, depend on coils and magnetic fields. Inductance is central to their operation.

The figure shows current and power waveforms for an ideal inductor supplied with a sinusoidal voltage, and for the moment we will focus on phases, assuming for convenience that the waveforms have unit amplitude. The key point to note is that the current waveform, coloured green, *lags* the voltage by a quarter of a period, in other words its phase angle is $-90°$ ($-\pi/2$ radians). Intuitively, this is because the magnetic field takes time to wax and wane, lagging behind the voltage waveform. Now $\sin(\omega t - \pi/2) = -\cos(\omega t)$, so the instantaneous power supplied to the inductor has the form:

$$p_L(t) = \sin(\omega t)\,(-\cos(\omega t)) = -\sin(\omega t)\,\cos(\omega t) = -0.5\,\sin(2\omega t)$$

$$(3.3)$$

This is shown as a red curve in Figure 3.4c. As with the resistor in Figure 3.4b, the power fluctuates at twice the supply frequency, but it now goes negative as well as positive. In other words the inductor alternately accepts power from the source and then returns it, so the average power is zero. It is not a net consumer of energy; rather it uses current from the source to establish an oscillating magnetic field.

To summarise, an 'ideal' inductor or coil, when supplied with a sine wave of voltage, draws a sinusoidal current that lags the voltage by 90°. The source, although supplying voltage and current, does not provide net power. The product of an inductor's voltage and current is, therefore, referred to as *reactive power* – one of the most important concepts in AC electricity. Reactive power is measured in *volt-amperes reactive (VAR)* to distinguish it from the *real power* dissipated by a resistor, which causes heating and is measured in *watts*.

So how is an AC generator affected by having to supply *reactive* power? Not much, you might think, because its main task is clearly to supply *real* power for heating or doing useful mechanical work. But there is a snag: although the current taken by a purely inductive load does not cause power consumption *in the load*, that same current must originate in the generator and flow along transmission lines and through transformer windings, all of which possess some electrical resistance and generate unwanted heat. So, in practice, a certain amount of energy is wasted getting reactive power to a reactive load, and power utilities tend to discourage reactive power demands which 'use up' a substantial portion of their equipment's current-carrying capacity.

And now for the last in our trio of circuit elements – the capacitor. In its simplest form a capacitor may be thought of as a pair of close-spaced metallic plates. It has the property of *capacitance*, measured in *farads* (F). We can deal with it relatively quickly because in many ways its behaviour is the exact counterpart of the inductor's. Whereas the AC current in an inductor *lags* the voltage by 90°, in a capacitor it *leads* by 90°. And whereas the inductor stores energy in a *magnetic* field, the capacitor stores it in an *electric* field set up between the plates. Like the inductor, the capacitor alternately accepts power from the source and then returns it, and the net power flow is zero. The current and power waveforms are shown in Figure 3.4d, and the instantaneous power takes the form:

$$p_c(t) = \sin(\omega t)\ \cos(\omega t) = 0.5\ \sin(2\omega t) \tag{3.4}$$

A capacitor also requires reactive rather than real power, but with phase relationships opposite to those of an inductor. Later we will meet several important ways in which capacitance affects AC power grids.

So far we have placed emphasis on phase relationships and the concept of reactive power. But how about the *amplitudes* of the currents that flow in inductors and capacitors? Specifically, since in their ideal form they contain no resistance, what limits the currents? The answer is their *reactance* which, like resistance, is measured in ohms (Ω); but, unlike resistance, reactance is frequency-dependent and involves the 90° phase shifts we have described above.

The reactance of an inductor is:

$$X_L = \omega L = 2\pi f L \quad \Omega \tag{3.5}$$

where L is the inductance measured in henrys (H). For example, in a 50 Hz AC system an inductor of 1 H has a reactance of $100\pi = 314\,\Omega$; in a 60 Hz system the same inductor has a reactance of $120\pi = 377\,\Omega$. The inductive reactance depends on the frequency.

The magnitude of AC current flowing in an inductor is simply equal to the AC voltage divided by the reactance. This is analogous to Ohm's law for a resistor – provided we remember the 90° phase shift. For example, a 1 H inductor, connected to a 50 Hz 220 V rms source, takes a current of $220/314 = 0.70$ A rms. The peak current is $0.70\sqrt{2} = 1.0$ A.

Whereas the reactance of an inductor increases with frequency, that of a capacitor reduces. It is given by:

$$X_C = 1/\omega C = 1/2\pi fC \tag{3.6}$$

Once again, the magnitude of AC current that flows is equal to the voltage divided by the reactance. As an example, you may like to check that a 0.0001 F capacitor connected to a 60 Hz 120 V rms source takes a current of 4.52 A rms.

So far we have considered resistors, inductors and capacitors as 'ideal' elements. In practice, of course, they are not ideal, for example, the wire of an inductor coil invariably has some resistance. In other cases, we need to analyse circuits containing separate R, L and C components. They all take sinusoidal currents when supplied with sinusoidal voltages, but amplitude and phase relationships are more complicated.

We can illustrate this with a simple example and at the same time introduce some further ideas and terminology. Figure 3.5a shows a resistor and inductor connected in series. The resistor's value is $R\,\Omega$, and we have labelled the inductor's reactance $j\omega L$, rather than just ωL. This simple change is in fact highly significant. The letter j (pure mathematicians use the letter i) denotes $\sqrt{-1}$, and denotes that the reactance involves a 90° phase shift. Mathematically, we are representing the reactance as an *imaginary* quantity, distinguishing it clearly from the resistance R which is represented as a *real* quantity. The j-*notation* allows linear AC circuits to be analysed simply and elegantly using the rules of *complex arithmetic*.

A full account of the j-notation applied to AC circuits is a standard ingredient of electrical engineering textbooks, but for our purposes it may be reduced to a few key ideas. To start, we note that the resistor and inductor

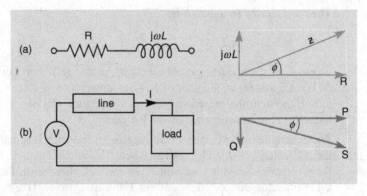

Figure 3.5 Simple AC circuits.

in Figure 3.5a are in series, and we may, therefore, add the resistance and reactance to give their combined *impedance*:

$$Z = R + j\,\omega L \tag{3.7}$$

The impedance, a complex quantity with real and imaginary parts, is represented in the figure. The resistance is drawn as a real number, the reactance as an imaginary number, and the impedance is found by vector addition. Alternatively, the impedance may be resolved into real and imaginary components. It is somewhat analogous to resolving the force acting on a turbine blade into axial and radial components – as we did in Figure 3.2.

The magnitude and phase of the impedance are given by:

$$Z = \sqrt{(R^2 + \omega^2 L^2)} \quad \text{and} \quad \varphi = \tan^{-1}(\omega L/R) \tag{3.8}$$

As an example, suppose that a coil with inductance 1 H and resistance 40 Ω is connected to a 50 Hz 220 V rms supply, and we wish to find the current that flows. The magnitude and phase of the impedance are:

$$Z = \sqrt{(40^2 + (100\pi)^2)} = 317 \ \Omega$$

$$\varphi = \tan^{-1}(100\pi/40) = 82.7° \tag{3.9}$$

The magnitude of the current is therefore $220/317 = 0.69$ A rms, and its phase angle with respect to the voltage is $-82.7°$. Note that since we are *dividing* the voltage by the impedance to find the current, the *positive* phase angle associated with the impedance produces a *negative* phase angle for the current, in line with the normal rules of complex arithmetic.

A similar approach may be used to represent the real and reactive components of the power supplied by an AC generator. Figure 3.5b shows a generator supplying voltage and current via a transmission line to a load. In general both line and load have reactance as well as resistance, and the generator's current is not in phase with its voltage. It supplies *real power* (*P*) to the resistive components of line and load, and *reactive power* (*Q*) to the inductive and capacitive components. The vector sum of these two powers gives the *apparent power* (*S*) supplied by the generator, equal to the product of its rms voltage and current and measured in *volt-amperes* (VA). For example, a generator might be supplying 100 kVA of apparent power, consisting of 80 kW of real power and 60 kVA of reactive power. Note that large quantities are expressed in kVA (kilovolt-ampere) or MVA (megavolt-ampere), equivalent to the use of kW and MW for real power.

As previously mentioned, the supply of reactive power uses up part of the current carrying capacity of a utility's generators and transmission lines and wastes real power in associated resistances. Commercial and industrial consumers requiring substantial amounts of reactive power are often charged special tariffs, depending on the *power factor* (*F*) of their loads. The power factor is defined as the ratio between real and apparent power, and from Figure 3.5b we see that it is also equal to the cosine of the phase angle φ. Thus:

$$F = P/S = \cos\varphi \qquad (3.10)$$

If the power factor is unity, no reactive power is supplied, the phase angle is zero, and the load is purely resistive. When reactive power is demanded, the power factor reduces and the phase angle increases. A positive phase angle means the load is capacitive, giving a *leading* power factor; a negative angle denotes an inductive load, producing a *lagging* power factor – typical of industrial loads that include large electric motors. A power factor of 0.9 is typical of the level below which utilities start to charge large consumers for reactive power. And generators, including marine energy devices, may be required to deliver their power at power factors close to unity.

The negative effects of reactive power may be counteracted, or offset, using *power factor correction* techniques. Imagine, for example that a load takes a current of 300 A at a lagging power factor of 0.8, equivalent to a phase angle of $-36.9°$. The current may be resolved into a real component of magnitude 300 cos 36.9° = 240 A, and a reactive component of magnitude 1000 sin 36.9° = 180 A. Now suppose capacitors, placed across the load, are designed to take a leading current of 180 A. This exactly counteracts the lagging current of the load and supplies the required amount of reactive power. The utility now provides just the real power via the transmission system, while the load and capacitor swap reactive power. This situation has, in fact, already been illustrated in Figure 3.4. Parts c and d show that the reactive current and power in the inductor and capacitor are in *antiphase* and will 'cancel each other out' if they have equal magnitudes. For this to happen their reactances must be equal at the supply frequency, so that:

$$\omega L = 1/\omega C \quad \text{and therefore} \quad C = 1/\omega^2 L \qquad (3.11)$$

A smaller (and therefore cheaper) capacitor may be used to give partial power factor correction – for example raising it from 0.8 to 0.9. Large industrial and commercial users of electricity often choose to install parallel banks of capacitors, switching in more or less capacitance automatically as the power factor of their load changes.

Utilities often use correction equipment of their own because power factors close to unity tend to enhance system stability as well as reducing resistive losses in generation and transmission. So far we have only mentioned capacitors; but inductors may be used to correct leading power factors, for example when feeding AC electricity into submarine cables possessing high capacitance.

Our final topic in this introduction to AC electricity concerns *three-phase systems*. So far we have dealt with a single sinusoidal voltage and the currents it produces in resistive and reactive loads. But, in practice, AC grids normally generate and transmit a set of three separate voltage *phases* (not to be confused with phase *angles*) displaced from one another by 120° ($2\pi/3$ radians). A set of such voltages is shown in Figure 3.6. Domestic customers with modest power requirements normally receive just one of the phases – a so-called *single-phase supply*. Commercial and industrial consumers are more likely to receive all three phases.

Three-phase AC has a number of advantages:

- Large three-phase generators, transformers and motors are more efficient than single-phase versions.

- Three-phase is more economical to transmit than single-phase, using less conductor material for the same power and voltage levels.

- Power flow from a three-phase generator into a 'balanced' linear load is constant (see below), reducing vibrations in the generator and three-phase motors supplied by it.

- Three-phase electricity can produce rotating magnetic fields, of major importance in the design of generators and motors.

Figure 3.7a shows a three-phase system comprising generator, transmission lines and load. The voltages for phases 1, 2 and 3 are produced by the generator's internal coils, or *windings*, which are arranged electrically as a 'Y' or 'star' and joined at the *neutral point* (N). Transmission is by three line conductors and one neutral conductor – a total of four cables. For simplicity we have not shown any transformers but these, too, have

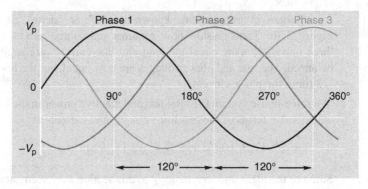

Figure 3.6 Three-phase voltages.

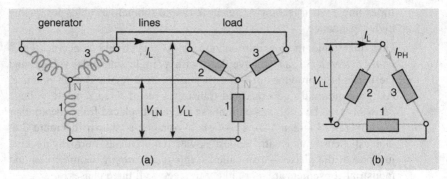

Figure 3.7 Three-phase AC power.

three-phase windings. At the receiving end the cables feed a three-phase load, such as a large electric motor. In addition, low-power single-phase loads, such as lighting, may be supplied by one line and the neutral cable, and distributed equally between the three phases. Households are normally supplied with one phase.

If the load is perfectly balanced with equal impedances in the three phases, the vector sum of currents at the neutral point is zero. In this situation the neutral conductor carries no current – which explains why it is normally much thinner and lighter than the three line conductors. In practice loads are unlikely to be perfectly balanced, so the neutral conductor carries a small amount of current.

The voltages of the various phases, measured between line and neutral and denoted by V_{LN}, are out of step by 120° – see Figure 3.6. The voltage V_{LL} between any two lines equals the vector sum of two phase voltages, giving $V_{LL} = \sqrt{3}\, V_{LN}$. The current in each line, I_L, equals the current in each phase, I_{PH}.

An alternative arrangement, known as 'Δ' or 'delta', is shown in Figure 3.7b. Transmission now requires only three cables, and if the three impedances are equal the load and line currents are again perfectly balanced. The line and phase voltages are now equal, but the line current is $\sqrt{3}$ times the phase current.

In a three-phase system the total real and reactive power in the load may be expressed in terms of the rms line voltages and currents as:

$$P = \sqrt{3}\, V_{LL}I_L \cos\varphi \quad \text{and} \quad Q = \sqrt{3}\, V_{LL}I_L \sin\varphi \tag{3.12}$$

Some of the ideas and terminology in this section may seem strange if you are new to AC electricity but they will be a great help for understanding the various types of generator used in marine energy devices, and the engineering challenge of integrating them into grid networks.

3.4 Generators

3.4.1 Introductory

It is often noted that the modern history of wind energy, a relatively mature technology, suggests many ways in which marine energy – and especially the technology of tidal turbines – is likely to progress and mature in the coming years. One of these ways concerns the choice of suitable electrical generators to feed renewable energy into AC grid networks. The account of generators given here therefore follows closely that in a recent book on wind energy by this author [2], in the expectation that what has worked for wind will work for wave and tide as the power ratings of individual devices reaches, and then exceeds, megawatt scale. We will start with a few historical notes and basic principles.

When Michael Faraday (1791–1867) discovered *electromagnetic induction* in 1832 he can have had little inkling that, within 100 years, the phenomenon would spawn a global electricity industry. A man of limited formal education, Faraday was appointed professor of chemistry at the Royal Institution in London in 1833 and is widely acknowledged as one of the greatest experimentalists of all time. He was fascinated by the relationship between electricity and magnetism and the possibility of making an electric motor. His theory of electromagnetic induction provided the essential insight; and today we realise that the fortuitous combination of Faraday's ideas, copper's conductivity, and iron's magnetism has formed the basis of electric motors and generators ever since.

From the point of view of electricity generation, the core principle of electromagnetic induction may be simply stated: a voltage is produced, or *induced*, in a conductor moving at right angles to a magnetic field. What matters is the *relative* movement – the conductor may be stationary and the field moving, or vice versa.

Figure 3.8 shows how this principle can be used to make a simple AC generator. A magnet with north and south poles rotates between the arms of a soft iron frame. Two slots in the frame carry an insulated copper wire in the form of a loop. The magnet forms the moving part – the *rotor* – and the frame and wire loop form the stationary part – the *stator*. As the magnetic field (green arrows) moves past the conductor equal but opposite voltages are induced in the two sides of the loop. Being in series they add together and appear at the terminals.

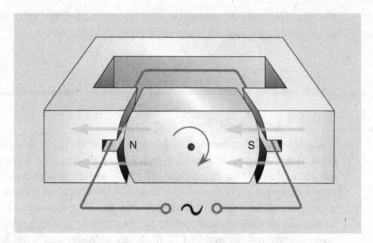

Figure 3.8 A simple AC generator.

Half a revolution later the north and south poles of the magnet have swapped positions, the field has changed direction, and the induced voltages change sign. This is an AC device that generates a fluctuating voltage – positive then negative – each time the magnet goes through a complete revolution.

In electromagnetic induction the magnetic field, the movement and the induced voltage are always at right angles, or *orthogonal*, to one another. Looking carefully at the rotational system of Figure 3.8, we see that in this case the field is *radial*, the movement is *tangential* and the conductor is arranged in *axial* slots.

How much voltage is generated as a magnetic field 'sweeps past' a conductor? This depends on two factors: the strength of the field and the speed of movement – in other words, on the *rate of change of magnetic flux*. The field may be maximised by keeping the 'magnetic resistance' or *reluctance* of the magnetic circuit as low as possible. In Figure 3.8 we have shown an iron frame that provides an 'easy', low-reluctance, path for the magnetic flux (apart from the small but necessary air gap between the frame and magnet). But even a strong magnetic field is unlikely to produce much voltage in a single conductor loop, so practical AC generators use coils with many turns of wire, referred to as windings. The instantaneous induced voltage in a winding is given by:

$$v(t) = -n \, (\mathrm{d}\Phi/\mathrm{d}t) \tag{3.13}$$

where n is the number of turns, Φ is the magnetic flux measured in *webers* and $\mathrm{d}\Phi/\mathrm{d}t$ represents its rate of change.

Although this simple machine demonstrates several key principles, it has its limitations. First, it is a single-phase device whereas large AC generators are normally three-phase. Secondly, its voltage waveform is unlikely to match the sinusoidal shapes shown in Figures 3.4 and 3.6. Although the magnet generates alternate positive and negative voltages as it rotates past the conductor slots, these are more likely to resemble 'pulses' than a smoothly-varying sine wave. In fact the production of sinusoidal voltages requires careful attention to the profile of the magnetic field and the layout of the copper windings, part of the stock-in-trade of the design engineer. Finally, the magnetic field of a large machine is often produced by windings on the rotor, supplied with DC, which acts as an electromagnet. This is referred to as *separate excitation*.

Figure 3.9a shows the main features of a high-power AC generator, this time producing three-phases. The windings are spread in slots around the stator's circumference. For simplicity we have shown them well separated, but they actually overlap and the number of conductors in each slot is carefully devised to produce sinusoidal voltage waveforms. As the rotor turns at constant speed the angular separation between the windings is translated into the required timing separation between the three voltage waveforms, shown in Fig. 3.9b. One complete AC cycle is generated for each complete revolution. If the required frequency is 50 Hz, the rotor must turn 50 times per second, or at 3000 rpm; if 60 Hz is required, at 3600 rpm.

So far we have assumed that the rotor has a single pair of poles, and this is generally true of large AC generators driven by steam turbines in conventional power plants, often with individual power ratings above 100 MW.

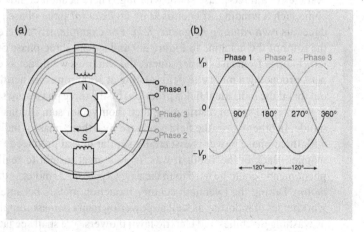

Figure 3.9 A three-phase AC generator.

Generators used in large wind turbines and marine energy devices, typically rated between one and a few megawatts, often have two pole pairs, halving the rotational speed for the same output frequency – for example 1500 rpm produces 50 Hz – with a corresponding reduction in the gearbox ratio. In principle, there is nothing against even more pole pairs, giving further speed reductions; indeed there is an important type of multi-pole, low-speed, generator that dispenses with a gearbox altogether, allowing it to be driven directly by a rotor – an option increasingly favoured for large wind turbines, including those deployed offshore. In general the speed of rotation in rpm is given by:

$$n = 60f/P \text{ giving } P = 60f/n \tag{3.14}$$

where f is the frequency of the generated supply and P is the number of pole pairs.

For reasons that will become clear shortly, the above machines are examples of *synchronous generators*, and the speed of rotation is known as the *synchronous speed*. When a machine's field is produced by separate excitation, it is known as a *wound rotor synchronous generator (WRSG)*; and when the field is produced by permanent magnets it is referred to as a *permanent magnet synchronous generator (PMSG)*.

There is a second major type of AC machine, of great importance for marine energy, known as the asynchronous or induction generator. Most people find it more difficult to understand than the synchronous generator, so it will be helpful to prepare the ground for a fuller description in later sections.

We have already seen how a rotating magnet generates a set of sinusoidal voltages in a three-phase stator winding. There is another side to the same coin: such a winding, if energised by an *external* three-phase supply, produces its own *rotating magnetic field*. For example, if we were to remove the rotor of the machine in Figure 3.9 and supply three-phase 50 Hz AC to its windings from an external source, the windings would act as electromagnets, producing a magnetic field rotating at 3000 rpm. Furthermore, it turns out that if the individual fields produced by the three phases are sinusoidal in both time and spatial distribution, then their vector sum is constant in magnitude. In other words the field produced by a three-phase winding supplied with three-phase power is essentially similar to that produced by a rotating magnet. Back in the 1880s it was Tesla's great insight to realise that this phenomenon could be used to make a new type of AC motor – the *induction motor*. During the twentieth century induction motors became the trusted workhorses of industry, as well as powering many domestic appliances such as washing machines and electric lawn mowers. We shall see later that their first cousins, induction generators, may be used to produce grid electricity.

Figure 3.10 Cyclists on a 'very long bike'.

What is the essential distinction between synchronous and asynchronous (induction) machines operating in a large electricity grid? A highly imaginative analogy based upon a number of cyclists riding a 'very long bike' has previously been described [1], and we will use a modified and simplified version of it here.

Figure 3.10 shows cyclists on a very long bike that represents the electricity grid. Some cyclists act as generators, supplying power to the grid. Others act as motors (loads), demanding power from the grid. Each has pedals connected to a sprocket wheel that meshes with a long chain running the whole length of the bike. The overall aim is to keep the bicycle moving along a straight flat road at constant speed, equivalent to maintaining the grid frequency at its nominal value of 50 (or 60) Hz. Air resistance and rolling resistance are neglected.

The 'generator' cyclists pedal actively, trying to push the bike forward, some with more force than others. The 'motor' cyclists apply braking force to the pedals, trying to slow the bike down. If the bike speed is to remain constant the total forces produced by the two groups must be kept in balance at all times. This is equivalent to a grid in which electricity consumption must always be matched by generation.

As well as being a generator or motor, each cyclist is either *synchronous* or *asynchronous*, depending on how his pedals are linked to the driving chain:

- *Synchronous.* The pedals of a synchronous cyclist are connected to the sprocket wheel by a stiff, but slightly elastic, shaft. If the cyclist is a generator, his effort twists the shaft slightly forward by an amount proportional to the torque exerted. But if the cyclist is a motor, the shaft is twisted slightly backward. As a result the angular positions of the various cyclists' pedals are not aligned – for example they do not all pass the lowest point at the same instant. The pedals of a generator are always slightly in advance of those of a motor. But it is important to realise that *synchronous cyclists must all pedal at the same speed*, dictated by the speed of the chain. In terms of our analogy, synchronous generators and motors connected to an electricity grid must rotate at a speed precisely determined by the grid frequency.

- *Asynchronous.* The pedals of an asynchronous cyclist are connected to the sprocket wheel by a *fluid coupling*. This consists of two small fan-like turbines facing each other in a fluid-filled housing. One turbine is rigidly connected to the pedals, the other to the sprocket wheel. The turbines are coupled together relatively 'softly' by the viscosity of the fluid, and as one rotates it tends to drag the other with it. A generator cyclist transmits power by pushing the pedal turbine faster than the sprocket wheel turbine; a motor cyclist does the opposite. Power transfer in either direction depends on a certain amount of *speed difference* or *slip* – no power is transmitted if the turbines are rotating at exactly the same speed. In terms of our analogy, an asynchronous generator must rotate somewhat faster, and an asynchronous motor somewhat slower, than the equivalent synchronous machine.

Let us now examine the cyclists in the figure more carefully. Suppose we are told that the four blue ones are all synchronous, and that one of them is acting as a generator and three as motors. Can we tell which is which? Indeed we can: cyclist 7 has his pedals slightly in advance of the others, so he must be the generator.

We are also told that the four red cyclists are asynchronous, with two generators and two motors. This time we cannot tell which is which because a 'snapshot' showing pedal positions gives us no information about rotation speeds and the amount of slip – the essential clue for distinguishing between generators and motors. We would need to know pedal *speeds*, not *angular displacements*.

The analogy offers further insights [1], especially into the question of reactive power balance. As we saw in the previous section, a generator or grid must generally supply two types of power: active power that delivers heat or mechanical work; and reactive power that supplies the out-of-phase AC currents demanded by inductive or capacitive loads. So far our bicycle analogy has focussed entirely on the real power generated or demanded by the cyclists. But the bike must also remain balanced and stable, not tipping over sideways and depositing the cyclists in the road. Forces trying to overturn the bike to the left may be considered analogous to producers of reactive power, for example capacitors used for power factor correction; forces acting to the right are analogous to consumers of reactive power, for example motor inductances. Note that the cyclists' balancing act involves constant adjustments of their weight to left or right but does not require them to generate (or consume) additional real power. This is analogous to a grid network, where reactive power balance is important for stability and maintaining the system voltage at the required level.

In the following sections we describe the main variants of synchronous and asynchronous generators. We will start with synchronous machines since they are in many ways easier to understand and relate to Figures 3.8 and 3.9. We will then move on to asynchronous machines which, together with recent developments in power electronics, offer attractive alternatives for generating grid electricity with marine energy devices.

3.4.2 Synchronous generators

Most conventional power plants produce electricity using synchronous generators driven by steam, gas or hydroelectric turbines. Two-pole machines coupled to high-speed steam turbines are often referred to as *turbo-alternators* and are accepted as trusted workhorses of the electrical supply industry. In Europe, with its 50 Hz grids, they rotate at 3000 rpm; in North America, with 60 Hz grids, at 3600 rpm. We have already illustrated their main features in Figure 3.9. Hydroelectric generators are generally designed as multi-pole machines, allowing rotation at the much lower speeds of large water turbines. But in all cases the generator is effectively tied to the grid it serves by a 'stiff magnetic spring', with no option but to rotate at *synchronous speed*. As its power output changes the rotor adopts a different *angle*, very much as the 'synchronous cyclists' described in the previous section did with their pedals; but it cannot alter its rotation *speed*.

Another important point is that, as we explained in the previous section, an electricity grid has to supply reactive as well as real power. One of the advantages of conventional synchronous generators is their ability to act as controlled sources of reactive power by varying the strength of their rotor fields, allowing them to run at either lagging or leading power factors. Such flexibility is a great asset for controlling and stabilising a large grid.

We may clarify the generation of real power by considering the interaction of magnetic fields. As already noted, the field of a wound rotor synchronous generator (WRSG) is produced by feeding DC excitation current into the rotor's field windings; a permanent magnet synchronous generator (PMSG) uses permanent magnets instead. In either case the rotor field rotates at synchronous speed and induces three-phase voltages in the stator windings (see Figure 3.9).

So far we have only mentioned induced voltages, but of course voltage on its own is not enough; the stator must also supply three-phase currents (and power) to the grid. This has an extremely important consequence: the currents produce their own rotating field that interacts with the field set up by the rotor. It is the 'tension' between the two fields, rotating synchronously but with an angular offset from one another, that produces torque and delivers real power to the grid. This is analogous to the stiff but slightly flexible shaft linking the pedals of a 'synchronous cyclist' to the chain of the bike shown in Figure 3.10.

Figure 3.11 illustrates the interaction of rotor and stator fields with a simple model. In this case fields are produced by permanent bar magnets, rotating together at synchronous speed n, but offset from each other by an angle β. North and south poles attract, so the angular offset produces a constant torque on the magnets that tries (but fails) to align them. In Figure 3.11a the torque on the rotor opposes its rotation, so the rotor must be supplying mechanical power and the machine is acting as a generator. In Figure 3.11b, with the offset angle reversed, the torque on the rotor aids its rotation, so the machine is acting as a motor. We see that positive values of β represent generation and negative values represent motoring – like most electrical machines, a synchronous generator can also act as a motor. In either case the rotor and stator are tied together by a 'magnetic spring'.

So what happens if a synchronous generator, coupled to a turbine via a suitable gearbox, is connected directly to a large electricity grid? The first point to note is that it is not self-starting and must be run up to synchronous speed before connection. The instant of connection must be finely judged, ensuring a suitable rotor angle as well as the correct speed. Thereafter,

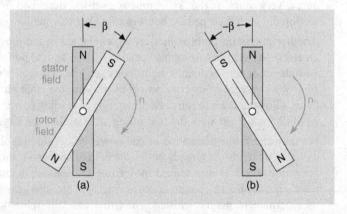

Figure 3.11 Interaction between stator and rotor fields (a) when generating and (b) when motoring.

the turbine must turn at a constant speed dictated by the grid frequency. If it tries to lag behind it begins to accept power from the grid, becoming a motor — hardly the desired effect!

There are two major reasons why synchronous generators are unsuitable for connecting wind turbines or marine energy devices directly to a grid:

- The connection is very 'stiff'. The grid demands a fixed rotation speed, allowing only small angular offsets of the generator's rotor. This is fine for synchronous machines driven by steam turbines, with power levels continuously under the control of supervisory engineers. But renewable energy sources tend to be highly variable and intermittent. If the generator's speed is fixed any power fluctuations translate directly into torque variations and power surges into the grid. A stiff, inflexible connection places high stresses on mechanical components and may induce vibrations. Even turbines equipped with variable-pitch blades may be unable to react sufficiently fast to counteract such effects.

- Turbine rotors operating in variable flows must be designed as variable-speed machines to achieve high efficiency and optimise annual electricity yields. This cannot be achieved with directly-connected synchronous generators.

At this point the prospects for synchronous generators may seem distinctly unpromising. But in this field of electrical engineering, as in many others, the situation has been transformed in recent years by impressive developments in *power electronics*.

If a synchronous generator driven by a variable-speed turbine cannot be connected *directly* to a large grid, perhaps it can be connected *indirectly* using a power-electronic interface? This would make the generator independent of grid frequency, voltage and phase. It may sound like an expensive option for a megawatt-scale device, but if the high cost of the electronics can be offset by gains in turbine efficiency, and possibly also by omitting a heavy and expensive gearbox, it can prove very attractive.

Figure 3.12 illustrates such a scheme for a tidal-stream turbine or other marine energy device. A synchronous generator is driven by the turbine via a gearbox. The gearbox is shown dotted to indicate that it can, in principle, be omitted if a multi-pole low-speed generator is used. The generator delivers power to an electronic *power converter*, also sometimes called a *frequency converter* because its key function is to decouple the rotation speed of the turbine from the frequency of the grid. This is done by first converting the generator's AC output to DC with a *rectifier*, then converting back again to AC using an *inverter*. The DC link provides isolation between the two unsynchronised AC systems, allowing the turbine to operate over a

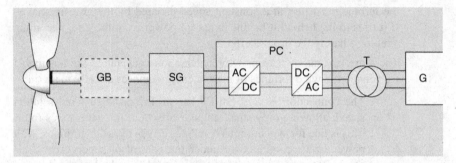

Figure 3.12 Main elements of a variable-speed turbine: GB = gearbox, SG = synchronous generator, PC = power converter, T = transformer and G = grid.

wide speed range. It also makes the grid connection much 'softer', reducing sudden surges of torque and electrical power as the tidal stream fluctuates. Finally, the AC output from the inverter is fed via a transformer into the grid. The scheme is referred to as *full-scale* or *fully-rated* power conversion because the full power output of the generator passes through the converter.

As well as allowing a generator to run at variable speeds, a power converter can be designed to produce or absorb reactive power, and to provide a 'soft start' for running the device gently up to speed. In principle, the scheme of Figure 3.12 may also be used with asynchronous generators because the power converter decouples the generated frequency from that of the grid. However, as we shall see later, asynchronous generators, especially large ones, are often used with *partial-scale* converters which handle only a portion of the output power and are considerably cheaper.

Rectifiers and inverters used in power converters are based on electronic switches, such as diodes, transistors and thyristors. Half a century ago the pioneers of solid-state electronics could hardly have imagined that semiconductor devices would one day control megawatts of power. But huge advances in power-handling capacity have been made since the 1980s, opening electrical engineering up to a wide range of new ideas and possibilities. Wind engineering has certainly been one of the beneficiaries, and we may expect marine energy to follow.

We see that a synchronous generator allied to a full-scale power converter provides decoupling of generator and grid frequencies, allowing a turbine or other device to work over a wide range of speeds, referred to as *full variable-speed operation*. In wind turbines a speed range between 2 : 1 and 3 : 1 is often chosen, sufficient to optimise energy capture over a wide range of operating conditions.

Multi-pole generators driven directly by turbine rotors without any need for gearboxes are also well-established in the wind industry. The main advantages are the elimination of a costly gearbox with its associated power losses; and the simplicity of the mechanical drive train, with a minimum number of bearings. However, low speed, high torque, generators are quite rare these days and demand special design and construction skills. They also have larger diameters than conventional machines, perhaps making them unsuitable for coupling to the relatively compact rotors of tidal-stream turbines.

In principle, the rotor fields of synchronous generators can be produced by permanent magnets rather than by separate-excitation. Permanent magnets avoid the power losses that arise when current is passed through copper windings, known as copper losses or i^2R *losses*, and can be made lighter and more compact. On the other hand the strength of the magnetic field is fixed, unlike that of a wound rotor which can be adjusted to control reactive power and output voltage – flexibility that is admittedly less important when the generator is connected to the grid via a power converter. These and other issues are well discussed elsewhere [1].

3.4.3 Asynchronous generators

3.4.3.1 Squirrel-cage and wound-rotor induction machines

Asynchronous generators connected to an electricity grid do not rotate at exactly synchronous speed because the generation of real power requires a certain amount of *slip*. Asynchronous operation is analogous to the action of a cyclist on a 'very long bike' whose pedals transmit power to the drive chain via a fluid coupling – a concept already discussed and illustrated in Figure 3.10. Turbines and other devices that drive asynchronous generators, also referred to as *induction generators*, are able to cushion sudden power fluctuations by speeding up or slowing down slightly, storing or releasing kinetic energy in the drive train and 'softening' the grid connection. Furthermore, as we shall see in the next section, recent developments in power converters allow induction generators to operate over a wide speed range.

In its basic form, an asynchronous (or induction) generator is cheaper and much simpler in construction than a synchronous machine with the same power rating. Yet the details of its operation are, by common consent, harder to visualise. Since most people are more familiar with motors than generators, and since the two are intimately related, we will start by explaining the action of an *induction motor*.

A three-phase induction motor has windings arranged around its stator in much the same way as a synchronous machine and, when supplied with a three-phase AC supply, produces a magnetic field that rotates at synchronous speed. It was Tesla's genius in the late 1800s to realise that

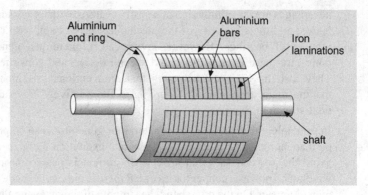

Figure 3.13 A squirrel-cage rotor.

the rotating field could be used to *induce* voltages in a simple form of rotor, and that the resulting currents would produce torque, tending to 'drag' the rotor around with the field. And so the induction motor was born.

The simplest type of rotor is known as a *squirrel-cage*, shown in its basic form in Figure 3.13. Iron laminations are built up to form a solid core, with a set of axial slots filled with cast aluminium bars that take the place of the more familiar (but more expensive) copper conductors. In practice, there are generally many more bars than shown in the figure, and they are skewed slightly to reduce torque fluctuations. The bars are short-circuited together by two *end rings* that complete the 'cage'. Squirrel-cage induction motors are produced in vast quantities for applications ranging from industrial drives to domestic appliances and have gained a reputation for robust reliability.

What happens when a stationary three-phase squirrel-cage motor is first connected to an AC supply? We may visualise the following sequence of events:

- A magnetic field, rotating at synchronous speed, is set up by the stator windings.
- Three-phase voltages are induced in the rotor bars, and since they are short-circuited by the end rings, currents flow. Initially the rotor is stationary, the currents have the same frequency as the supply, and the rotor produces its own magnetic field that also rotates at synchronous speed.
- Interaction between the stator and rotor fields produces torque and the rotor starts to turn. As it gathers speed, attempting to 'catch up' with the stator field, the frequency of the induced currents in the rotor bars decreases.

- The rotor can never quite reach synchronous speed. If it did, there would be no induced currents in the rotor bars, no torque, and the machine could not act as a motor delivering mechanical power. For torque and power to be produced, there must be a certain amount of *slip*. This contrasts with a synchronous machine, which rotates at precisely synchronous speed.

We see that the key to motor operation is the *induction* of rotor currents by the stator's magnetic field, which tries to 'drag' the rotor with it. The slip is defined as:

$$s = (n_s - n)/n_s \tag{3.15}$$

where n is the rotor speed and n_s is the synchronous speed. Slip is generally quoted as a percentage of the synchronous speed, with values between about 0.5 and 3% typical for motors when delivering their rated power. Large high-power motors generally run at lower slip values than small ones.

As an example, suppose a large three-phase two-pole induction motor connected to a 50 Hz supply is running at 1% slip. Its rotation speed is 99% of synchronous speed, in other words $0.99 \times 3000 = 2970$ rpm. Note that the rotor is 'slipping back' through the stator field at 30 rpm or 0.5 revolutions per second. It follows that the frequency of the induced rotor currents has fallen right down to 0.5 Hz, and that the field they produce is rotating at just 30 rpm *relative to the rotor*. But since the rotor is itself turning at 2970 rpm, the field is rotating at 3000 rpm *relative to the stator*. The same argument holds for any other value of slip; the stator field and induced rotor field both rotate synchronously, interacting to produce a steady torque, even though the rotor body is moving at less than synchronous speed.

So how does an induction machine work as a generator? The answer is 'rather easily'. As a general principle, rotating electrical machines can act either as motors or generators. If they accept electrical power and deliver torque, they are motoring; but if they accept torque and produce electricity, they are generating. We should also note that, to a reasonable approximation, induction machines may be considered *linear*. An induction motor running just below synchronous speed can be 'pushed' up to and beyond synchronous speed by applying external torque to its shaft. Above synchronous speed the slip and torque simply change sign and the machine starts to generate. There is nothing to prevent a smooth transition from motor to generator.

Figure 3.14 shows how the torque of a typical induction machine varies as it works up from zero speed (*start-up*) towards synchronous speed as a motor, and becomes a generator above synchronous speed. Speed values

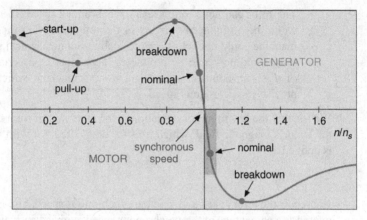

Figure 3.14 Typical torque–speed characteristic of an induction machine.

are normalised to the synchronous speed n_s. The motor portion of the characteristic, shaded blue, is important for turbines that use their generators to provide motorised run-up, and we may identify several important points on the curve:

- *Start-up.* The torque and currents at zero speed tend to be large, well above the nominal rating of the machine. Therefore, rather than connect the machine suddenly to the supply, a controller, known as a soft-starter, is normally used to ensure a gentle start-up.

- *Pull-up.* The pull-up, or minimum torque must comfortably exceed the resisting torque for the run-up to continue.

- *Breakdown.* The breakdown torque is the peak value available when the machine is running as a motor with low values of slip. If exceeded, the motor stalls.

- *Nominal.* The nominal torque is the value at which the machine is designed to operate for long periods.

Following a motorised run-up a turbine can start generating. Torque and slip change sign, and the generator runs slightly above synchronous speed. For example, if the slip is −2%, the normalised speed $n/n_s = 1.02$. In normal operation torque fluctuations produced by the turbine cause small changes in slip. Note that only a very small portion of the complete generator characteristic is used during normal operation – indicated by dark green shading in the figure. The machine generates within a narrow speed range – so much so that a turbine of this type is often described, rather confusingly, as *fixed speed*.

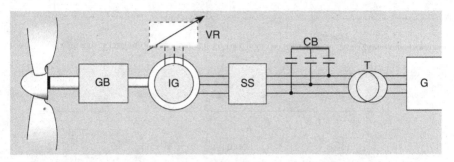

Figure 3.15 Main elements of a 'fixed speed' turbine: GB = gearbox, IG = induction generator, VR = variable resistance, SS = soft starter, CB = capacitor bank, T = transformer and G = grid.

Figure 3.15 illustrates the main elements of such a turbine. The rotor and gearbox drive an induction generator, shown by two circles representing its stator and rotor (for the moment we will ignore the optional variable resistance connected to the rotor). Three-phase stator windings feed electricity to the grid via a transformer. A soft-starter is included to aid motorised run-up. Power-factor correction is provided by a capacitor bank, a point that deserves some explanation.

One of the most important features of induction machines is that (not surprisingly) they are highly inductive, requiring reactive power to build up and maintain their magnetic fields. Whereas synchronous generators produce their own fields and have inherent flexibility for controlling voltage and reactive power, an induction generator always runs at a lagging power factor and must be supplied with reactive power by the grid, or by a local capacitor bank. From a circuit point of view its stator and rotor possess both resistance and inductance, and the interaction of the various elements as speed varies determines the shape of the torque–speed characteristic. When we recall that inductive reactance is proportional to frequency, and that the frequency of induced rotor currents is proportional to the amount of slip, it is not surprising that the characteristic shown in Figure 3.14 has such a complicated form.

In any case it is important to realise that Figure 3.14 is typical rather than definitive. In particular, it can be modified by altering the amount of rotor resistance. A squirrel-cage rotor, with its cast bars and end rings, has fixed resistance; but an alternative form of rotor uses copper windings instead of aluminium bars, producing a *wound rotor induction generator* (WRIG). The ends of the windings are brought out to *slip rings* on the rotor shaft, and electrical contact is made by *carbon brushes* that bear upon the slip rings (note that the word *slip* is here being used in a different sense). Although slip rings and brushes make WRIGs less robust than simple squirrel-cage machines, a wound rotor introduces some very important

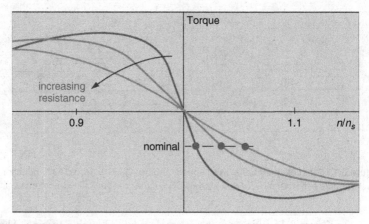

Figure 3.16 Typical effects of increasing the rotor resistance of a wound-rotor induction machine.

options – including the addition of variable resistance to the rotor, shown by the dotted lines in Figure 3.15.

Typical effects of adding external resistance are shown in Figure 3.16, which focuses on the region close to synchronous speed that is relevant to a machine working as a generator. The main effect is that the slip for a given output torque (and power) increases as the rotor resistance increases. Starting with the variable resistance set to zero (red curve), the rotor windings are effectively short-circuited and the rotor behaves like a squirrel-cage. Adding resistance (orange and green curves) reduces the slope of the torque-speed characteristic, increasing the operating speed range from, say, 1% to 5% of synchronous speed. A given torque fluctuation produces a larger speed change. This allows sudden power surges to be partially absorbed by changes in turbine speed, producing a 'softer' grid connection and reducing mechanical stresses on turbine components. It is sometimes referred to as *high-slip* operation.

The addition of rotor resistance causes power losses, which would normally be considered a disadvantage. However, if a turbine rotor is producing excess shaft power, the introduction of rotor resistance allows the excess to be dissipated in the rotor circuit, keeping the stator power within its rated value. This is a valuable alternative to shedding power by pitching the blades, which may be too slow in turbulent conditions.

A further option provided by a wound rotor is known as *slip power recovery* [1]. The power losses incurred by adding rotor resistance reduce the generator's overall conversion efficiency – and the higher the slip, the greater the losses. It would be attractive if, instead of dissipating power as unwanted heat, slip power could be 'recovered' and put to good use.

Any AC power taken from the rotor is at variable frequency and cannot be fed directly into the grid. However, it can be rectified and then inverted by a power converter to make it grid-compatible, adding to the flexibility of high-slip operation.

Thirty years ago it was common practice to design wind turbines with directly-coupled asynchronous generators and fixed blades, giving robust and relatively inexpensive 'fixed speed' machines. Unable to vary their speed by more than a small percentage, or to pitch their blades to accommodate changes in wind speed, such machines were relatively inefficient at converting wind power to electricity, and relied on *passive stall* to protect them from damage in storm conditions. Today the market for large wind turbines has swung decisively towards variable-speed machines and pitched blades – using either synchronous generators and full-scale power converters as described in the previous section or, as we shall see in the following pages, induction generators with full or partial-scale converters. It will be interesting to see what effect these developments have on the marine energy market in the next decade – and especially on the design of tidal stream turbines.

3.4.3.2 Doubly-fed induction generators

In the previous section we shifted attention gradually from the simplest form of induction machine with its squirrel-cage rotor to wound rotor machines. In particular we saw how a wound rotor induction generator is not restricted to accepting mechanical power from the rotor and delivering electrical power to the stator via rotating magnetic fields. Electrical power may also be *extracted* from the rotor via slip rings and brushes, modifying the torque–speed characteristic. We now take this important idea a stage further by explaining how electrical power may also be *fed into* the rotor, producing one of the most important recent advances in variable-speed turbine technology – the *doubly-fed induction generator* (DFIG) [4].

The DFIG concept has several key advantages:

- It allows turbines to operate over speed ranges up to about 2.5 : 1, sufficient to give high overall efficiency. This is sometimes referred to as *limited variable-speed* operation, distinguishing it from the fuller variable-speed operation of synchronous generators.

- It incorporates a *partial-scale* power converter which is considerably cheaper, and has lower losses, than a full-scale converter.

- The power converter may be designed to include facilities such as soft-start, voltage and torque control and power-factor correction.

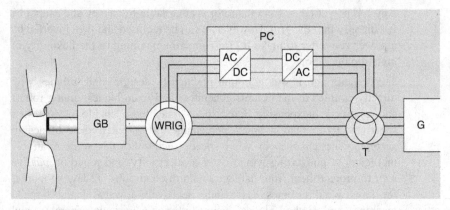

Figure 3.17 Main elements of a variable-speed turbine using a doubly-fed induction generator: GB = gearbox, WRIG = wound-rotor induction generator, PC = power converter, T = transformer and G = grid.

For all these reasons DFIGs have been increasingly favoured as the generators for large wind turbines in recent years, a trend that may well transfer over to tidal stream turbines and other marine energy devices as they attain, and exceed, megawatt power ratings.

Figure 3.17 shows the main components of a DFIG scheme. A turbine and gearbox drive the shaft of a wound rotor induction generator. The generator's stator is connected to the grid via a transformer. In addition its rotor can supply electrical power to, or receive power from, the grid via a partial-scale power converter. The term *doubly-fed* refers to the fact that the stator and rotor can both be supplied with AC voltages. In this arrangement the power converter must clearly be able to transfer power in either direction, in other words it is *bi-directional*.

If you are familiar with conventional electrical generators, this probably seems a curious system. A generator normally takes in mechanical power on the rotor and delivers electrical power from the stator. The idea – even the possibility – of the rotor accepting and delivering *electrical* power seems strange to most people. So we will start with a few general observations.

The generator at the centre of the scheme has two three-phase windings, one on the rotor and one on the stator. In principle, both are able to generate rotating magnetic fields and transfer power across the air gap. The machine's characteristics depend on the *relative* speed difference between rotor and stator – the amount of slip. It is even possible to imagine clamping the rotor and allowing the stator to spin! All this points to an essential equivalence between stator and rotor. A stator normally delivers real power, and accepts reactive power, from the grid. But there is no reason, in principle, why the

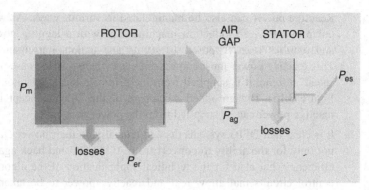

Figure 3.18 The flow of real power in a doubly-fed induction generator.

rotor should not also send or receive electrical power, using brushes and slip rings on its shaft, even if it is turning and receiving mechanical power from the outside world at the same time. This insight, and its practical application, give extraordinary flexibility to the turbine designer, producing cost-effective machines that adapt their speed to fluctuating energy flows.

So where does the power captured by the turbine actually go? The *power balance* in Figure 3.18 illustrates the flow of real (but not reactive) power through the system, assuming electrical power is being delivered to the grid by the rotor as well as the stator. On the left-hand side, the input mechanical power to the rotor is denoted by P_m. The rotor delivers electrical power P_{er} to the grid (via the power converter). There are small power losses in the rotor, due mainly to resistance in its windings, and the residual input power is transferred across the air gap to the stator. There are also small stator losses, and the remainder is delivered to the grid as stator electrical power P_{es}.

The alternative is for electrical power to be *fed into* the rotor by the power converter. The arrow denoting rotor power now points the other way, but the power balance of the system must still be maintained. For example, the extra power might be used to increase the torque and/or speed of the turbine rotor, or it might be fed across the air gap to the stator.

But how can electrical power be fed into the rotor? In the previous section we saw that power can be taken out of a wound rotor via slip rings and brushes, and dissipated as heat in a variable resistor. That may seem fairly straightforward, but feeding power in is harder to visualise, especially since the frequency of voltages and currents in the rotor varies with the amount of slip. The situation may be summarised by saying that power can be 'pushed back' provided the power converter injects AC voltage into the rotor at the right frequency, with appropriate amplitude and phase. This is certainly possible using today's sophisticated power electronics.

Reactive power can also be manipulated in various ways. As we pointed out previously, an induction machine runs with a lagging power factor and requires reactive power to set up and maintain its magnetic field. This reactive power must come from somewhere – in the case of a 'fixed-speed' turbine, it is supplied from the grid or a local capacitor bank (see Figure 3.15). But an added advantage of the DFIG concept is that the reactive power can be supplied from the power converter.

It is clear that DFIG systems depend crucially on their power converters – not only for the ability to convert from AC to DC and back again at high efficiency, but also for the technical sophistication of the algorithms and control circuits that allow real and reactive power to be taken from, or sent to, the rotor. As Figure 3.17 indicates, this requires two bi-directional AC/DC units (rectifier-inverters) together with a DC link (shown by red and blue lines) that decouples the frequency of the rotor from that of the grid and allows the turbine to run at variable-speed.

The AC/DC unit connected to the rotor is known as the *rotor-side converter*; that connected to the transformer and grid is known as the *network-side converter*. They are controlled independently of one another. Although there is considerable flexibility when allocating roles to the two units, in many cases the rotor-side converter provides torque control and voltage or power-factor control for the DFIG; the network-side converter controls the voltage of the DC link and transfers rotor power to and from the AC system at unity power factor.

The DFIG concept offers a comprehensive range of control options, but understanding the details requires expertise in electrical machines and circuit theory beyond the scope of this book. In the rest of this section we focus on some operational aspects of DFIGs, as they affect the turbines that drive them.

As already noted, probably the most valuable attribute of the DFIG is its variable speed. A rotor speed range up to about 2.5 : 1 is normally sufficient to give good turbine efficiency over a wide range of operating conditions. This is less than the 3 : 1 range typical of synchronous generators used with full-scale power converters, but it is considerably cheaper. It turns out that a DFIG can produce the required speed variation using a partial-scale converter rated at about one third of the maximum power produced by the turbine. Not only is this a lot cheaper than a full-scale converter, it also saves power losses in the converter itself.

The speed variation is obtained partly by operating the DFIG *super-synchronously*, in other words above synchronous speed, as with the 'fixed speed' squirrel-cage and wound rotor machines already described; and partly by operating it *sub-synchronously*, or below synchronous

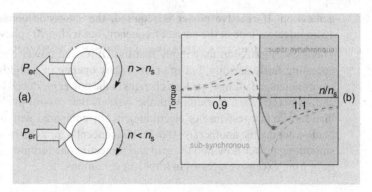

Figure 3.19 Sub and super-synchronous operation of a DFIG.

speed. Sub-synchronous power generation may seem surprising because we have previously assumed that an induction machine acts as a motor below synchronous speed (with positive slip), and as a generator above synchronous speed (with negative slip). But in fact a DFIG can operate as a generator both sub- and super-synchronously. Basically this is because electrical power may be fed into, as well as out of, the rotor. In either case the stator feeds energy into the grid.

The situation is summarised by Figure 3.19a. At the top, real electrical power P_{er} is taken out of the rotor, which runs above synchronous speed ($n > n_s$). But electrical power may also be supplied to the rotor, which then runs at less than synchronous speed ($n < n_s$). The desired speed ratio is achieved by a seamless transition from sub- to super-synchronous operation. Typically, in a megawatt device the speed can be made to vary between about 60 and 140% of synchronous speed using a partial-scale power converter able to handle about a third of the full rated power of the turbine. This corresponds to a range of slip values between plus and minus 40%, far greater than with conventional induction generators.

Figure 3.19b shows how the torque–speed characteristic of a typical induction machine begins to change as real power is fed in to or out of the rotor. In a DFIG system this is achieved by injecting a voltage with suitable frequency and phase from the rotor-side converter. Positive torque corresponds to motoring, negative torque to generation. The red curve shows the characteristic of the basic machine, which changes from motoring to generation as the speed passes through synchronous speed ($n/n_s = 1.0$). The yellow curve shows what happens with negative voltage injection; the machine only starts to generate at super-synchronous speeds. Conversely, with positive voltage injection, indicated by the orange curve, generation starts at sub-synchronous speeds. The solid portion of each curve between the two coloured dots indicates the region for real power

161

generation. If reactive power is required, the rotor-side converter must again inject voltage at the correct frequency, but with a 90° phase shift.

The above discussion may imply that the DFIG and power converter are operating fairly steadily, but in practice they operate in a highly dynamic environment, adapting to sudden changes in the power produced by the rotor and, in many cases, working with the systems that control pitching of the blades. The fast response of electronic circuits compared with mechanical blade-pitching is another of the DFIG concept's principal advantages, smoothing power flow into the grid and protecting the turbine and its drive train from excessive stresses in turbulent conditions.

3.4.4 Linear motion generators

The vast majority of electric generators are *rotational*, equipped with a central shaft and rotor which delivers power across a narrow air gap to a circular stator – a very natural arrangement. But what if the primary mechanical motion is linear or *translational*, as in a point-absorber WEC that heaves up and down with the waves? Rather than generate electricity via an intermediate hydraulic or pneumatic stage, as described in Section 3.2, or via connecting rods and crankshafts as in conventional diesel engines, is it possible (and sensible) to convert linear motion directly into electricity using a novel form of generator? This question has exercised the minds of wave energy experts for many years and various schemes have been suggested [3]. We focus here on one that is easy to visualise and illustrates the key principles.

Figure 3.20a shows a heaving float that drives a linear generator fixed to the sea bed. The float's motion is directly coupled to a *translator* which moves up and down between the arms of a stator, generating electricity. We should emphasise straight away that, for simplicity, we have omitted the generator housing, bearings and other important details. A housing is of course necessary to protect the unit and allows the translator to be separated from the stator by a small air gap.

Figure 3.20b shows a section through part of the generator. On the left, copper conductors (C) form a set of three-phase windings imbedded in slots in the steel of the stator. Next comes a small air gap (A), followed by the translator which carries permanent magnets (M), shown in green, which are arranged with alternating polarities. The magnets produce a powerful field that 'circulates' flux through translator and stator, as indicated by the red arrows. As the translator moves up and down the conductors cut the lines of magnetic flux and deliver a three-phase voltage to the stator terminals. If a load current is taken from the terminals, the stator produces its own magnetic field which reacts with that of the translator to produce mechanical force. The principles involved are exactly the same as for a

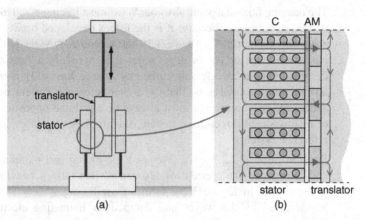

Figure 3.20 A linear motion generator.

permanent magnet synchronous generator (see Section 3.4.2), except that the motion is now linear rather than rotational. In effect the linear generator behaves like a rotational generator in which the rotor and stator have been 'laid out flat'.

The most obvious advantage of such a *direct-drive* scheme is its mechanical simplicity. But linear motion generators have their own special problems and design challenges:

- The magnetic circuit is 'open' at both ends, unlike that of a rotational machine which forms a closed circle. The degree of magnetic interaction between the translator and stator varies as the translator moves up and down, especially near the ends of the stroke.

- The stroke length, typically up to a few metres, is only optimal over a narrow range of wave heights.

- The machine is essentially a very low-speed, high-force device compared with conventional rotational generators which run at high speed and low torque.

- The power output depends on the instantaneous speed of the translator, which varies continuously as it follows the waves. This produces an 'unruly' electrical output, highly variable in magnitude and frequency, which can only be grid-connected via a sophisticated electronic power converter.

- Short-term overloads can produce severe mechanical and electrical stresses, especially since the direct-drive approach includes no intermediate energy storage. Such machines may be especially vulnerable in extreme wave conditions.

163

The heaving float of a point-absorber WEC must be connected to a reference system of some sort because it is the forces developed between them that drive the generator. In Figure 3.20 the reference system is the sea bed, but it need not be so; in deep water it is normally a floating structure designed to be relatively undisturbed by surface waves [3]. Newton's third law of motion assures us that, whatever reference system is employed, action and reaction forces must always be equal and opposite. In the case of a slow-speed direct-drive system, this implies very large forces on the reference system.

Back in Section 2.2.3.2 we discussed the tuning and damping of WECs and introduced the concept of resonance. Basically a heaving float and generator, as in Figure 3.20, form an oscillating system whose motion is energised by the waves and damped by extracting electrical power from the generator. The degrees of tuning and damping are crucial to the performance of the device and, in particular, to its energy absorption and efficiency. Damping may also be crucial for limiting the translator's stroke length and speed in extreme wave conditions, ensuring survival. All these considerations pose major design challenges.

Our coverage of basic AC electricity in Section 3.3 is helpful for understanding the electrical aspects of linear motion generators. First, we should note that the term *linear generator* has rather different meanings for mechanical and electrical engineers. To the first, it implies motion in a straight line; their electrical colleagues, however, are likely to interpret it as a generator whose performance can be modelled by a set of linear circuit elements, including resistors and inductors. Actually a synchronous generator like that in Figure 3.20 is linear in both senses – or sufficiently so to justify modelling each phase of its three-phase output by the simple *equivalent circuit* shown in Figure 3.21. We now discuss each circuit element in turn:

- *Voltage source (V)*. This represents the AC voltage generated in the stator as the translator oscillates up and down. As noted above, it

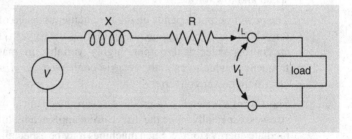

Figure 3.21 Equivalent circuit of one phase of a synchronous generator.

is highly variable in both magnitude and frequency. It is referred to as the *open-circuit voltage* because it is the value measured at the generator terminals when no load is connected. As we pointed out in Section 3.4.1 (see also Equation 3.13), the magnitude of the induced voltage depends on the strength of the magnetic field, the speed of movement, and the number of turns in the stator winding. Since the maximum speed of the translator is typically just 1 or 2 m s^{-1} – around 10 times lower than the equivalent in a conventional generator – it is important to produce a strong field. Modern permanent magnets based on rare-earth elements are particularly effective in this role.

- *Reactance X.* The reactance of the stator winding is referred to as the *synchronous reactance*. Its value is not constant, but varies throughout the stroke as the frequency of the generated voltage fluctuates. It has two components, known as the *main inductance* and the *leakage inductance*. The former, the larger portion, is crucial to the operation of the generator since it represents magnetic coupling between the stator and translator fields; the latter represents unwanted flux leakage. If a resistive load is connected to the generator and a current I_L flows, the voltage drop across the synchronous reactance reduces the terminal voltage below its open-circuit value V.

- *Resistance R.* This circuit element represents the resistance of the stator winding, which also reduces the generator's terminal voltage when current is drawn, and gives rise to real power losses in the stator. In principle R may be reduced by increasing the diameter of the stator's copper conductors. However, this reduces the cross-sectional area available for the steel teeth that provide a low-reluctance path for the magnetic flux, so a compromise is necessary.

- *Load.* When a load is connected a current I_L is drawn and the voltage reduces from V to V_L. In practice, the 'unruly' current produced by a large linear motion generator must be rectified (AC to DC) and changed back again to AC with the correct frequency and phase by a power converter before connecting to an electricity grid – a process similar to that for the variable-speed tidal turbine illustrated in Figure 3.12. The power converter can also control the load's power factor, provide phase compensation and regulate the power taken by the load, offering a valuable control strategy for protecting the WEC against overloads.

The electrical properties of a linear motion generator, including its synchronous reactance and resistance, have a profound influence on the properties of the WEC as a whole. Electrical and mechanical properties constantly interact and successful design must ensure not only efficiency in converting the highly variable energy of waves into grid-compatible electricity, but also survivability in extreme conditions.

3.5   Connecting to the grid

3.5.1   Setting the scene

The story of grid electricity goes back over a century. Edison's pioneering work in the late 1800s on the generation and distribution of DC electricity was overtaken in the twentieth century by the AC systems championed by Nikola Tesla (see Section 3.3), largely due to the ease with which AC voltage can be transformed – up for long distance transmission, down for local distribution and consumption. Early grids, often owned by private companies or municipalities, were generally small and local but were soon being interconnected to form larger systems. Governments in industrial countries became keen to encourage the growth of electricity, realising its importance for industrial innovation and efficiency and its key role for enhancing their citizens' quality of life. Today it is impossible to imagine a modern economy and society without large amounts of electrical energy at its core.

As grids expanded, gaining regional or even national coverage, the size of individual power plants increased dramatically. In the 1930s the largest were typically rated at around 60 MW; half a century later they were exceeding 1 GW. Currently the largest UK power plant is the Drax station in Yorkshire, producing up to 4 GW from six steam turbine generators each rated at 660 MW. The remarkable story of electricity supply in the twentieth century is one of ever larger centralised power plants based on fossil fuel combustion and, in some countries, nuclear fission and hydroelectric generation. By the end of the century many people had come to see this as an established, stable, recipe for the future – but their cosy view was already being disturbed by the arrival of significant amounts of renewable energy. To put this into perspective it is helpful to contrast the geographical distribution of conventional power plants with regions most favoured with abundant renewable energy.

A good example of an extensive system serving a developed economy is the UK's national grid, originally developed in the 1930s to form Europe's largest synchronised AC electricity network. Operating at 50 Hz and up to 132 kV, it was upgraded in 1949 with some 275 kV links, followed by additional 400 kV links from 1965 onwards. The highest voltages are used to transmit three-phase bulk electricity around the country, transformed down to progressively lower voltages for distribution to industrial, commercial and domestic customers. There are also submarine links for bulk power transfer to and from France (2 GW) and The Netherlands (1 GW).

The map of England, Wales and Scotland in Figure 3.22 shows the current locations of conventional power plants rated above 500 MW. These are principally coal-fired stations in traditional industrial areas; modern gas-fired units more widely dispersed; and nuclear plants, generally away from highly populated areas and close to the coast. Many smaller plants are not shown in the figure; nor is the vast distribution network ranging from high-power, high-voltage cables supported on huge pylons right down to the 230 V cables that bring single-phase AC electricity into individual households.

The UK's pattern of large power plants underlines a number of economic and operational criteria that apply equally well in other countries:

- Power plants are best located close to centres of population and industry, reducing the need for long transmission lines and minimising transmission losses.

- Fuel transport costs are reduced by siting plants close to fuel sources (or coastal terminals if the fuel is imported).

- Power plants needing large quantities of cooling water are often placed beside shorelines, large lakes or rivers.

- Public acceptability demands that nuclear plants are generally located well away from population centres.

Such considerations have largely determined the distribution of power plants shown in Figure 3.22. But what about renewable energy, and are the rules of the game being altered by its rapid growth? Will grid networks need substantial modification, and must network designers and operators adopt new mindsets and learn new skills?

Nature does not generally cooperate by providing renewable energy where it is most needed. We are now in an era of increasing renewable generation, with grid penetration levels of 10%, 20% or even higher on the horizon. In some European countries wind and solar energy are already contributing substantially to electricity supplies, and we may reasonably expect marine energy to join the club in the next decade or two. Clearly, such plans have major implications for grid operators who must allow for increasing amounts of variable generation while maintaining security of supply. Big challenges undoubtedly lie ahead for electricity supply industries in many developed countries – probably the biggest since the arrival of nuclear power over half a century ago.

Regions favoured with abundant renewable energy do not generally meet the criteria for siting grid-connected power plants. In the UK, wind farms

Figure 3.22 The distribution of large conventional power plants in the UK.

are often found in remote hilly areas of Scotland and Wales, or offshore in the North Sea. The county of Cornwall has an abundance of wind, solar and wave energy. In Scotland, the Orkney Islands and Western Isles, including Islay, offer huge potential for marine energy. But, as Figure 3.22 shows, these areas lie well away from the main centres of population and heavy investment in new transmission systems is needed if they are to realise their full potential. Nor is it simply a matter of grid *connection*, important though that is; it is also one of *integration* – the successful blending of renewable and conventional energy sources into a large grid network to provide safe, reliable and economic electricity on a national, or even international, scale. Hidden from view and attracting little public attention, grid integration is one of the biggest challenges facing the global electricity industry.

3.5.2   Grid strength and fault levels

Electricity grids are generally thought of as 'one-way' systems. Large AC generators in power plants (preferably out of sight, usually out of mind) feed electricity into the grid for delivery to homes, offices and factories. Like water and gas supplies, the flow is one way. Who can imagine pumping water back into their kitchen taps, to be credited by the water company; or making their own gas, and pumping any excess back into the gas main? Yet in recent years it has been possible for households to generate renewable electricity from wind, sun or a flowing stream and export the electricity back into the cables that normally import it. In principle, an electricity grid can be supplied or tapped at any point.

Of course a domestic installation with a peak power output of a few kilowatts is a very different proposition from a wave or tidal machine rated at 1 MW. Nobody is suggesting connecting a 1 MW device, still less a 10 MW array, to the domestic electricity supply. But suppose an array of tidal stream turbines is to be grid-connected in a sparsely populated area. For a start it is obviously necessary to lay cables to the nearest suitable connection point. But what is needed to make an efficient connection that is acceptable to the grid operator and local consumers? In this section we deal with two important requirements:

- Marine energy devices, when operating at maximum output, must not overload local distribution cables and transformers.
- Their variability, connection and disconnection must not cause unacceptable voltage fluctuations in the electricity supplied to local consumers.

To explore these issues it is first necessary to appreciate the electrical layout of a typical grid, which is illustrated in Figure 3.23a. On the left, large power plant generators (we show just two, G_1 and G_2), generally with individual ratings well above 100 MW, feed three-phase electricity into a high-voltage (hv) transmission grid via step-up transformers (T_1 and T_2). Power is tapped at various points and fed into a medium-voltage (mv) distribution system via a step-down transformer (T_3) for delivery to local areas. Finally it is transformed down again (T_4) to a voltage suitable for individual consumers (office, home, street lights, etc.), who generally receive just one phase of the three-phase supply. Larger consumers, such

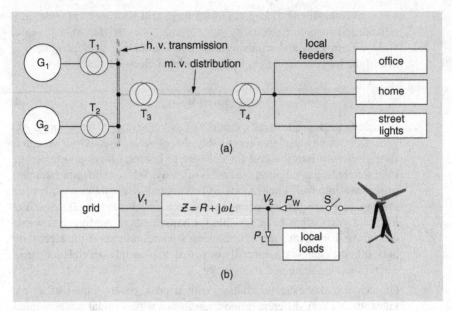

Figure 3.23 (a) A typical grid network and (b) a tidal turbine connection.

as factories, are often connected to the mv system and supplied with three phases. The voltage levels in various parts of the system vary from country to country, but are typically between 110 and 750 kV for hv transmission, 10 and 70 kV for mv distribution and 120 or 240 V for final delivery to consumers. We have shown transformers in the figure but not the associated switchgear, fuses and circuits that control power flow and prevent damage under fault conditions.

All this is conventional 'one-way' delivery from large centralised power plants to millions of individual consumers. There is no hint so far of renewable energy being fed into the grid, often by relatively small generators in far-flung parts of the network. To discuss this we need some basic knowledge of AC circuits and associated terminology, and you may wish to refer back to the introduction given in Section 3.3.

Figure 3.23b illustrates, in basic form, the electrical connection of a large tidal turbine (or other marine energy device) to the mv distribution network. We assume that the grid on the left includes the large synchronous generators, transformers and hv cables that make up a conventional high-power generation and transmission system, delivering a stable three-phase AC voltage (V_1) at a frequency of 50 or 60 Hz to the mv distribution network, represented by an impedance Z. On the right is the turbine with a rated power P_W together with switchgear (S) that connects it to the nearest suitable point of the network. Also connected at this point are existing local loads that consume power (P_L). The voltage at the connection point is V_2.

We start with the tidal turbine disconnected. In this case the local loads draw power from the grid in the normal manner. Ideally, the voltage V_2 they actually receive is very close to the nominal system voltage V_1. However, the current taken by the loads must flow through distribution cables and transformers, which have a combined impedance (Z) made up of resistance (R) and inductive reactance ($j\omega L$) in series. This causes a voltage drop equal to the product of current and impedance magnitudes, and has the effect of reducing the voltage V_2 below V_1. Clearly, if the impedance is too large, or if too much current is being drawn, the voltage drop may become unacceptable. This is especially likely to happen if large loads are connected at the far end of a long distribution line.

You may be surprised at the representation of all supply cables and transformers by a single impedance Z. Typically, distribution cables are mainly resistive, causing real power losses; transformers are very efficient at transferring real power, but their windings are inherently inductive and need reactive power to set up and maintain their magnetic fields. So the combination of cables and transformers in a distribution network normally produces a complicated mix of resistance and reactance. Fortunately, AC circuit theory comes to the rescue, allowing us to lump all the individual elements together and represent them by a single impedance.

This idea has important implications for describing the quality of a grid connection. If Z is small (cables and transformers generously rated) in relation to the currents demanded by the loads, the voltage V_2 received by the loads remains very close to the nominal system voltage V_1 and the grid connection is said to be *strong* or *stiff*. But if the impedance is comparatively large (low-rated cables and transformers) V_2 may deviate substantially from the system voltage and the grid connection is said to be *weak*. Furthermore, the voltage will fluctuate substantially as individual loads are switched on and off – not a good recipe for customer satisfaction. Needless to say, strong grids are highly desirable from the customers' viewpoint and grid operators normally undertake to maintain supply voltages within, say, 10% of their nominal values. But generous cables and transformers are expensive to install, especially in outlying areas, and a strong grid connection can easily become a weak one if more customers and loads are connected.

The usual way of quantifying grid strength at a particular point is by its *fault level*. This is the product of the nominal system voltage and the current that would flow in the event of a short-circuit. In Figure 3.23b the short-circuit current would be V_1/Z, so the fault level in VA is $V_1{}^2/Z$. For example, if the nominal voltage of a distribution system is 30 kV and it has an impedance of 9 Ω at the connection point, the fault level is equal to:

$$(30 \times 10^3)^2/9 = 10^8 \text{ VA} = 100 \text{ MVA} \tag{3.16}$$

The fault level gives a good indication of the maximum load that can be connected without causing an unacceptable voltage drop. By definition the short-circuit current is the maximum that could possibly flow, and it reduces the voltage V_2 to zero. But when a normal load is connected the acceptable voltage drop is far smaller – say 5%. This in turn implies that the load VA should not exceed a few percent of the fault level. For example, loads totalling up to 5 MVA could probably be connected satisfactorily to a point where the fault level is 100 MVA.

We now come to an important question: what happens when the tidal turbine is connected to the grid and starts generating? We are assuming in Figure 3.23b that the turbine power P_W is connected to the grid at the same point as some local loads. Energy cannot be stored, so the power flows from the turbine, into the loads, and through the impedance Z must always be in balance. Furthermore *apparent power*, equal to the product of AC voltage and current and measured in VA, comprises *real power* measured in watts (W) and *reactive power* measured in VAR – and both types of power must remain in balance.

All this leads to a complicated power balance at the grid connection point, making it difficult to predict exactly what will happen to the voltage V_2 as the turbine's power output varies and local loads are switched on and off. Generally, we may expect that whenever $P_W > P_L$ the voltage V_2 will tend to rise above V_1 as the turbine 'pushes' real power into the grid through impedance Z; but when tidal generation falls below local load demand, V_2 will fall below V_1. Detailed analysis is beyond the scope of this chapter, but it is clear that a strong grid with low impedance Z is highly desirable because it 'ties' the voltage at the connection point more firmly to the nominal system voltage V_1, preventing unacceptable fluctuations as the turbine output and load demand vary. Large turbines attached to weak grids are a recipe for trouble!

We see that fault levels are good indicators of a grid's ability to accept power from a marine device or array. For example, if an array of 10 devices with a rated output of 10 MW is to be installed, it would be wise to seek a grid connection with a fault level above, say, 100 MVA. If this is not available a network upgrade may be necessary. A key issue for grid operators as penetration levels of renewable energy continue to grow is the provision of grid capacity with adequate fault levels at all points of connection.

Such issues have been extensively studied in the case of wind energy [1], and we may expect the lessons learned to benefit the new marine energy industry. For example, weak grids caused major problems in California during the early days of the modern wind energy renaissance, sometimes referred to as the 'Californian wind rush'. The Tehachapi Pass in the mountains north of Los Angeles had a huge number of wind turbines, relatively small by modern standards, connected to an existing grid network. Unfortunately, the 66 kV grid had been designed to serve a population of less than 20 000 people spread over a wide area, with a planned load of just 80 MW, but by 1986 the power output of Tehachapi turbines was peaking at 125 MW and the system became overwhelmed. If ever there was an example of a weak grid, this was surely it. At the time there was very little understanding of the problem, which was exacerbated by the use of conventional induction generators needing large amounts of reactive power that simply could not be transported over the cable network. Such mistakes, it seems safe to say, would not be repeated today, for wind or other renewable technologies. Actually the Tehachapi problem was far greater than the connection of individual turbines or wind farms; it affected the operation and stability of the grid as a whole, and the lessons painfully learned were essentially those of *grid integration* – a theme we will develop a little later.

3.5.3 Electrical quality

In the previous section we focussed on the voltage fluctuations that affect an electrical grid, particularly when the grid is weak at the point of connection. The problem is important because electrical equipment, from industrial motors right down to consumer electronics, is designed to work at particular voltages and grid operators aim to keep voltage levels reasonably steady. However, there is more to 'electrical quality' than relatively slow changes in grid voltage over time scales of minutes or hours. The *rapidity* of fluctuations is also very important.

Tidal turbine and wave energy devices can, like wind turbines, produce rapid voltage changes that make lights flicker and annoy nearby consumers. Such *voltage flicker* may be caused by stream turbulence, the pulsating nature of wave energy, or device connection and disconnection. It is important to realise that the amount of flicker caused by a large device depends greatly on its hydraulic and electrical design. For example, 'fixed speed' turbines based on induction generators are particularly susceptible, whereas today's variable-speed machines tend to iron out flicker effects by absorbing power surges as changes in rotor kinetic energy. Electronic power converters offer great scope for rapid and flexible control of the voltage and power delivered to a grid.

The human eye is most sensitive to brightness fluctuations at frequencies of around 8–10 Hz, but flicker comes in many guises and is highly subjective. Occasional connection and disconnection of megawatt devices may be acceptable whereas repeated on–off cycles can try the patience. Random flicker caused by turbulence is perceived as qualitatively different from regular, periodic, flicker and all are highly dependent on the strength of the grid at the point of connection. This makes any annoyance very hard to predict or quantify.

Another important aspect of a grid's electrical quality concerns waveform distortion. AC voltages at all points on a grid are ideally sinusoidal in form, and so are the currents drawn by *linear* loads, as assumed in our introduction to AC circuits in Section 3.3. But *nonlinear* loads take non-sinusoidal currents from the supply. A familiar example is a dimmer switch controlling the brightness of an electric light, which works by switching the current rapidly on and off during each cycle of the voltage waveform. Dimmer lights are hardly likely to upset the operation of a grid, but large nonlinear loads are a different matter. By demanding non-sinusoidal currents they tend to distort the voltage waveforms appearing on a grid – especially where it is weak – and in serious cases may produce overheating or failure of other equipment, faulty operation of protective devices, nuisance tripping of sensitive loads and interference with communications circuits. Electric utilities normally impose strict limits on the amount of voltage distortion that can be introduced into their networks by large nonlinear loads.

The other side of the coin concerns generation. The large synchronous generators that feed conventional electricity grids are expertly designed to produce sinusoidal voltage waveforms with minimal distortion, but marine energy devices may be far less ideal. Electronic power converters incorporate high-speed switching circuits that, by their very nature, can only approximate smooth sinusoidal waveforms. Here is a considerable challenge for designers, who must ensure that any distortion remains within acceptable limits.

Waveform distortion is normally assessed in terms of *harmonics*. A pure sinusoidal voltage contains a single frequency, referred to as the *fundamental*. But if the waveform becomes distorted additional frequencies, known as harmonics, are introduced at integer multiples of the fundamental. For example, a distorted 50 Hz waveform may contain harmonic components at 100, 150, 200 Hz and so on. Switching waveforms tend to be rich in harmonics. The underlying concept is that of *Fourier analysis*, named after French mathematician and physicist Jean Baptiste, Baron de Fourier (1768–1830) who showed that any periodic waveform, regardless of its shape, may be formed by adding together a set of sinusoids with appropriate amplitudes and phases. Fourier analysis crops up widely in science and engineering, and we have already met it in our discussion of wave spectra in Section 2.1.3.

The amount of distortion in a waveform is usually measured in terms of *total harmonic distortion* (THD), which is the ratio of the energy in all harmonic frequencies to the energy in the fundamental, expressed as a percentage. THD in the high-power transmission section of a large electrical grid is normally very small – say below 1% but it tends to grow as the electricity is transformed down and distributed to local areas and, finally, to individual consumers. Large tidal turbines and WECs incorporating power converters will tend to raise the THD on the distribution network, especially if the grid connection is relatively weak.

We have concentrated on the potential effects of marine energy devices on the quality of grid electricity, and for anyone interested in their design and operation these are clearly important issues; yet there is another side to the story – the effects that fluctuations of grid voltage and frequency, and grid outages, may have on the operation and safety of devices. Not surprisingly, these are likely to be more serious in a weak grid and, once again, operational experience in the wind industry may prove valuable; countries as diverse as Germany and India, with large numbers of wind turbines connected to mv distribution systems, have encountered various problems caused by two-way interaction between grids and turbines [1].

In this section we have summarised a number of relatively local issues that may arise when marine and other renewable energy devices are connected to an electricity grid. It is now time to consider the effects that renewable energy has on grid systems as a whole – and how they may be expected to cope with increasing amounts of installed capacity.

3.6 Large-scale renewable energy

3.6.1 Introductory

In previous sections we discussed local problems that may arise when connecting marine energy devices to a grid, and the importance of grid strength at the point of connection. Local voltage fluctuations, flicker and waveform distortion are certainly important to local consumers, but they hardly ruffle the feathers of a large grid. What matters to system planners and operators is the reliable and economic supply of electricity over a regional, national or even international, network.

A large modern electricity grid connects hundreds of generators to thousands of kilometres of transmission and distribution lines and finally to millions of consumers. As marine energy starts to make a significant contribution, substantial upgrades to existing cable networks will be needed. Good examples are provided by the Scottish islands, including Orkney and the Western Isles, which are powerhouses of wave and tidal stream energy. But their full potential will only be realised when expensive new submarine cables are installed linking them to the mainland. Successful grid integration, as well as local connection, are key issues for device developers.

Grid integration of increasing amounts of wind power has received much international attention over the past 30 years [1]. More recently, solar photovoltaic (PV) power has come of age, reaching gigawatts of installed capacity in many countries, including Germany, Spain, Italy and the USA. Enthusiasts of marine energy naturally hope that the next 20 years will see wave and tidal stream devices join the family of renewable technologies and make significant contributions to national electricity supplies. Successful grid integration will become, in many countries, a question of accepting substantial amounts of power from a number of renewable technologies with different operational characteristics.

Conventional grids are powered by synchronous generators driven by steam, gas and hydroelectric turbines that are firmly 'tied' to the grid and must rotate at speeds dictated by the grid frequency. Individual generators are able to control their terminal voltage and power factor by varying field excitation. Operators keep the system frequency and voltage within narrow limits by matching total generation of real and reactive power to total demand, requesting changes in output from various power plants in a highly skilled balancing act.

Renewable electricity tends to disturb the complex, but well-established, control of conventional synchronous generation in several ways:

- It is intermittent, on different time scales, and (with the exception of tidal power) relatively unpredictable.
- It uses a variety of generating systems, synchronous and asynchronous, and electronic power converters.
- The rated power of individual devices and small-to-medium arrays or 'farms' (say 1–20 MW) is much smaller than that of single, large conventional generators (say 100–750 MW).

None of these differences is of much concern to grid operators as long as the penetration of renewable energy remains small. But as it rises, and as the capacity of individual installations begins to rival that of large conventional power plants, the picture changes and successful grid integration moves to the top of the agenda. It is time to consider some large-scale system issues that affect the operation of the grid as a whole.

3.6.2 Intermittency and variability

Ocean waves wax and wane; tidal streams ebb and flow. The outputs of devices that generate electricity from them are inherently intermittent and variable, unlike the large conventional power plants that have dominated the electricity supply industry for over 100 years. However it is important to realise that the short-term variability of an individual device is generally greater than that of a multiple array. In the case of a wind farm, different wind conditions across the site mean that some turbines may be generating while others are stationary. When all are working, some generate more power than others and individual outputs vary in response to wind turbulence. Since the fluctuations are at least partly unsynchronised (uncorrelated) between different turbines the result is reduced variability in the output of the farm as a whole. This important effect, which grows with the capacity and extent of a wind farm, is referred to as *power smoothing*. We may expect similar effects in arrays of wave and tidal stream devices.

In any case, the operators of large electricity grids are certainly not concerned with the intermittency and variability of individual devices, or even arrays – unless they are very large. More significant is the *total* generation at a given time, gathered across the network from widely dispersed installations experiencing different meteorological and ocean conditions. It is rather misleading to describe renewable energy as intermittent at the power system level. It is certainly variable, but the power smoothing effect is greatly enhanced by a mix of technologies and a good geographical spread of installations. For example, wind, wave and tidal conditions in the west of Scotland hardly ever coincide; nor do they match those in the south of England. In future there will be trading of renewable power between nations, leading to even more 'geographical smoothing'. But the big picture is rarely understood, or appreciated, by casual observers of marine energy devices or idle wind turbines.

In the modern development of renewable energy, wind power may be said to have really 'taken off' in the 1980s and solar PV in the 2000s. Wave and tidal stream power may join forces in the 2020s. In the future, countries with a good mix of technologies, spread over the widest areas, will be best placed to integrate renewable energy into their grid networks.

It is also very important to realise that conventional generating plant does not operate continuously, nor is it 100% reliable at times of peak demand. In fact the public might be quite surprised to learn how often conventional plant is shut down in fault conditions or for maintenance – and how long repair can take, especially in the case of nuclear power where safety is such an issue.

So far we have concentrated on the *supply* side of electricity generation, but an equally important aspect is *demand*. In the absence of energy storage, stable operation of an electricity grid requires supply and demand to be kept in balance at all times and this greatly affects any discussion of intermittency and variability. Figure 3.24 shows a record of supply and demand over a

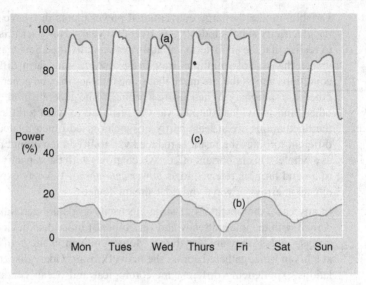

Figure 3.24 Electricity demand and supply on a large grid.

seven-day period in a large grid with a substantial percentage of renewable generation, expressed as percentage of peak demand. The demand curve (a) displays a fairly regular daily cycle with a morning peak as the population rises and starts work, a flattening off in the middle of the day, and a smaller peak in the early evening. Loads are smaller at the weekend because industry and commerce are largely inactive. Renewable generation, shown by curve (b), has peaks and troughs dictated by nature rather than by human activity. In this example it supplies up to about 20% of peak demand but not, of course, in synchronism with it. The shape of the curve depends on the relative contributions of the various renewable technologies. For example, a large PV contribution would tend to produce visible peaks around the middle of each day; a substantial tidal stream contribution might introduce peaks at roughly 6-hourly intervals day and night, and so on. The difference between the red and blue curves, labelled (c), represents the power that must be supplied on a minute-by-minute basis by all non-renewable generation such as fossil-fuel, hydro and nuclear.

Curve (a) is crucial to our discussion because it shows that grid operators are already used to high levels of variability, not in supply, but *in demand*. They must cope with large daily swings, often up to 40 or 50% of peak demand, and schedule generation capacity accordingly. Conventional power plants are continually being brought in and out of service, adjusting their outputs, and matching demand on a minute-by-minute basis. Actually, consumer demand can be even more fickle than suggested by the regular peaks and troughs in the figure. Famous – we might say notorious – examples are the sudden surges that occur at half-time in a World Cup football match, or during breaks in popular TV dramas, when people watching television rush to their electric kettles to brew quick cups of tea or coffee. The effect, known as *TV pickup*, is well known to the UK's national grid operators who have learned to cope with gigawatt fluctuations within a few minutes, balancing the sudden 'ramping' of demand with generation by fast-response power plant. The speed of such changes rivals anything that waves and tides can do!

We see that variability is a major and inherent feature of large electricity grids. Demand fluctuates on a wide range of time scales, from minutes to months and even years. On the supply side, faults occasionally shut down large power plants or transmission lines. The injection of renewable power certainly adds to the overall variability of the system, but not to the extent that many people assume, and the effect will reduce as renewable portfolios diversify both technologically and geographically.

Successful adaptation to the variability of renewable power is made much easier by accurate forecasting. Tidal streams are some of the most predictable flows in nature; and great advances have been made over the past 20 years in wind and wave forecasting, not least because of the value of accurate predictions to wind farm operators and fishing fleets.

The central message is that intermittency and variability are not serious problems for large electricity grids unless caused by the sudden, unpredictable, shut down of large power plants. Renewable generation simply makes a contribution to the highly variable picture already produced by outages of conventional plant and fluctuations in consumer demand. That there are no insuperable technical problems, at least up to penetration levels of around 20%, seems proved by the experience of pioneering countries such as Germany and Spain that have already integrated impressive amounts of wind and solar energy into their grid networks.

3.6.3 Capacity credit and backup generation

From an environmental point of view the most obvious benefit of renewable energy is that it is 'carbon free'. A unit of electricity generated by wind, wave or tide obviates the need to generate it by other means, including the burning of coal, gas or oil. For this reason renewable generation is often described as a 'fuel saver' that reduces carbon dioxide emissions. So far, so good – but what about its ability to displace other types of generation? This is a key question because all types of power plant are expensive to build and maintain. For over 30 years sceptics have repeated the claim that, since the output of wind turbines is intermittent, wind is an inherently unreliable energy source that cannot reduce the need for conventional generating capacity. They maintain that it is simply a fuel saver, without *capacity credit*; and that there will always be need for *backup generation*. The same criticism will no doubt be levelled at marine energy as it begins to contribute significantly to national electricity supplies.

All this relates to our discussion in the previous section. If you monitor the power output of an individual WEC or tidal turbine, or even an array of devices, you may conclude that marine energy is highly variable. But grid operators are concerned with total generation and regard renewable energy as one more variable input to the system. In their eyes all types of generation and consumption are, to some extent, 'unreliable'.

Back in 2006 the UK's Energy Research Centre (UKERC) produced a detailed report [5] based on a review of more than 200 international studies. None suggested that renewable energy generation at levels up to 20% of demand would compromise the reliability of the British electricity system over the following 20 years, although it might lead to some increase in costs and would affect both short-term management and long-term planning.

Short-term management of a grid requires operators to *balance* electricity generation and consumption continuously, scheduling power plants over time scales from minutes to hours in order to keep the system frequency and voltage within statutory limits. Operators must be able to call on *balancing reserves* that can boost generation at short notice. The introduction of substantial amounts of renewable power means that the output of other plant will need to be adjusted more frequently, and the UKERC report found that a certain amount of backup generation – in the case of wind, about 5–10% of installed capacity – may be needed to cope with peaks in demand that are uncorrelated with peaks in renewable generation.

An important feature of balancing reserves is their speed of response, or *ramp rate*. Clearly, a sudden increase in electricity consumption (we mentioned the 'World Cup problem' in the previous section), or a sudden loss of output from other generators due to fault conditions, cannot be offset by reserves that take hours to come online. Fast-response generators with high ramp rates, such as hydroelectric and gas turbines, can work up to full output within minutes and are said to be *dispatchable*. The outputs from nuclear and, to a lesser extent, coal-fired power plants take far longer to adjust and for this reason they are normally used to supply the system's minimum demand level, or *base load*, typically around 50% of the peak (see Figure 3.24).

In addition the ramp rates associated with a particular type of reserve capacity are affected by whether or not it is kept *spinning*. Synchronous generators that are 'up-to-speed' and locked to grid frequency do not have to generate power. They can just spin at synchronous speed, consuming minimal fuel and reacting quickly when power is demanded. Conversely *non-spinning* reserves must be brought up to speed before synchronising with the grid. In the case of a steam turbine this is best done gently to avoid severe thermal stresses as it approaches working temperature.

Grid operators normally accept renewable power whenever it is generated, because it is a question of 'use it or lose it'. This means, of course, that the output from other generators must be adjusted accordingly. The ramp rates of reserve capacity must be able to cope with the fastest rates of change in renewable generation – caused, for example when a storm sweeps rapidly across the North Sea. However, this does not mean that dedicated reserve capacity need be provided to support individual installations; indeed it would be very uneconomic to do so. The variability of wind, wave and tidal generation, and the backup generation needed to support them, is only significant at the system level.

To summarise the short-term management issue, successful balancing of supply and demand in the presence of renewable generation does require some extra backup plant, but the balancing act is no different in principle to that already required by other types of generation and by the highly variable nature of consumer demand.

The second main impact of renewable energy on a large grid network concerns medium to long-term system reliability. Planners aim for a safety margin in total grid capacity compared with anticipated peak loads, allowing for unexpected surges in demand and forced outages of large power plants. Unlike short-term system balancing, they must often look 15 or 20 years ahead, taking into account predicted growth in demand and the retirement of plants reaching the end of their working lives. The *system margin*, also called the *planning reserve*, is essentially concerned with strategic planning of generating capacity to ensure security of supply in the future – in popular parlance, 'to prevent the lights going out'.

No electricity system is 100% reliable because there is always a small chance of failures in power plants and transmission networks, especially when demand is high. Even a substantial system margin cannot give an absolute guarantee of reliability. There is, very roughly, a 10% chance of a conventional power plant, fossil-fuel or nuclear, being out of action at any particular moment and in a worst-case scenario the loss of a major plant can produce a domino effect on the whole system. For example, in 2003 there was a massive blackout in the northeastern USA caused by cascading power outages – and it had nothing to do with renewable energy. In 2008 seven large UK power plants, including one nuclear station, unexpectedly shut down causing widespread power cuts described by one industry executive as a 'gigantic coincidence' – which had nothing to do with renewable energy – and the same nuclear plant was offline for more than six months in 2010. The tragic accident at Japan's Fukushima nuclear plant in 2011 showed the whole world just how unpredictable and dangerous major power plants can be. In the midst of such risks and uncertainties the job of planners is to try and provide a margin that represents a sensible compromise between system reliability and cost, taking account of the existing and anticipated generation mix, including renewables.

The *capacity credit* for any type of generation is, effectively, a measure of the contribution it can make to system reliability, and is normally expressed as a percentage of installed capacity. In the case of wind this figure is predicted by the UKERC report to be approximately 20–30% in British conditions, assuming generation reaches 20% of total demand. It is clear that wave and tidal stream capacity cannot be expected to displace conventional generation on a one-for-one basis, either for short-term system balancing or for longer-term system reliability. But taking an overall view, the international consensus among today's planners and system engineers is that large modern grid networks can accept 20% or more of diverse renewable generation without serious technical or operational problems; and that any difficulties arising tend to be managerial and organisational, requiring a shift of mind set away from the large centralised generation of the twentieth century towards something more fragmented and flexible in the twenty-first.

If this sounds optimistic and reassuring, a lingering doubt must remain about how much capacity credit should be given to renewable generation, and what backup should be provided to cope with exceptional circumstances. As noted above, no grid system can ever be 100% reliable and tomorrow's marine energy will simply add its own element of uncertainty. Efficient planning may increase reliability to a very high level, but it can never achieve certainty – that is in the nature of engineering systems. So we will end with a story, an imagined scenario that might conceivably occur in the not-too-distant future.

> By the 2040s the countries of northwest Europe, linked by a North Sea Supergrid, have become used to trading offshore wind and marine energy generated over the sea's vast stretches. It is rare for winds, waves and tides to die simultaneously but in the freezing winter of 2045 a freak high pressure system settles over northwest Europe for two weeks, virtually killing all winds and waves. At the same time the hydroelectric reservoirs of Norway and Scotland have become unusually depleted; five large UK nuclear power plants are shut down for safety checks; and a trade dispute has severely reduced imported supplies of gas and oil from the Middle East.

On 14th February vast numbers of Europeans decide to watch a momentous international football match on TV, and at half-time 20 million households switch on electric kettles to brew tea and coffee. The lights (and the TVs) go out – and do not come back on again for some considerable time.

Fact or fantasy – a miniscule risk to be ignored, or a significant one to be planned for? How often will extreme meteorological events occur, and how long will they last? These are big unanswered questions for renewable energy, and only time will tell. In the interim we would surely be wise to understate the capacity credit of renewable generation and diversify our energy supplies as much as possible, accepting that no sources can ever be completely reliable. And maybe, by 2045, all European consumers will be connected to electricity grids by very smart meters, utilities will vary the cost of electricity hour by hour to control demand, and international football matches will have gone out of fashion.

References

1. T. Ackermann, ed. *Wind Power in Power Systems*, John Wiley & Sons, Ltd: Chichester (2005).
2. P.A. Lynn. *Onshore and Offshore Wind Energy: An Introduction*, John Wiley & Sons, Ltd: Chichester (2011).
3. J. Cruz, ed. *Ocean Wave Energy: Current Status and Future Perspectives*. Springer: Berlin (2008).
4. B. Wu, Y. Lang, N. Zargari, S. Kouro. *Power Conversion and Control of Wind Energy Systems*. Wiley-IEEE Press: New York (2011).
5. R. Gross and P. Heptonstall. *The Costs and Impacts of Intermittency: An Assessment of the Evidence on the Costs and Impacts of Intermittent Generation on the British Electricity Network*, UK Energy Research Centre (UKERC), March 2006.

4 Case studies: Wave energy converters

4.1 Introductory

The practical development of wave energy converters (WECs) has reached a critical stage and there is a realistic prospect that, within the next few years, various designs will progress from full-scale prototypes to commercial deployment. Forty years on from the pioneering work of Professor Stephen Salter, the wave energy community is seeing its dreams realised. In this chapter we describe a selection of devices that have reached, or seem very likely to reach, megawatt-scale. It is perhaps unlikely that all will reach full commercialisation because there are many hurdles, financial and environmental as well as technical, to be jumped; but a wide-ranging selection has the advantage of illustrating the dynamic nature of wave energy research, the many approaches being adopted to translate the wild motion of ocean waves into well-behaved grid electricity, and the hopes of professional engineers in many countries to realise their visions.

It is hard to detect any overarching principles in WEC design beyond the need to extract energy from waves as efficiently and economically as possible and ensure survivability in extreme conditions. Section 2.2.2 mentioned one widely-used classification system, recommended by the European Marine Energy Centre (EMEC) in Orkney, which divides today's WECs into seven categories based on their scientific principles:

- Attenuator
- Oscillating wave surge converter

Electricity from Wave and Tide: An Introduction to Marine Energy, First Edition. Paul A. Lynn.
© 2014 John Wiley & Sons, Ltd. Published 2014 by John Wiley & Sons, Ltd.

- Oscillating water column
- Overtopping device
- Point absorber
- Submerged pressure differential
- Other.

In the following pages we will meet examples in various categories, many of which have been, or are being, tested at EMEC. After a brief historical introduction and indication of current status, we focus on the technical features and principles underlying each device, particularly as they relate to earlier chapters.

4.2 Case studies

4.2.1 Pelamis

Pelamis Wave Power Ltd [1, 2] was formed in Edinburgh in 1998 to develop a unique design of WEC which has since become one of the most researched and tested of all WECs [3]. By 2003 an extensive programme of tank testing using models between 80th and 7th scale had been carried out at universities in Edinburgh, Glasgow, and London, and also at facilities in Trondheim, Norway and Nantes, France. In 2004 the first full-scale prototype, the *Pelamis P1*, was installed at EMEC [4] in Orkney where it became the first offshore WEC to supply electricity to the UK's national grid. Following modifications in Edinburgh the improved machine formed the basis of a joint project between Pelamis and a Portuguese utility to set up the world's first experimental wave energy farm in 2008. Three devices with a combined capacity of 2.25 MW were installed at Aguçadoura, 5 km off the northwest coast of Portugal. Although the global financial crisis of that year precipitated the collapse of the utility's parent company, a great deal of valuable operational experience had been obtained and was put to good use in a second prototype – the *Pelamis P2*.

By 2010 orders had been secured from utilities for two of the new P2s – one each for E.ON and ScottishPower Renewables – with an agreement to test them in tandem at EMEC to gain as much performance data as possible. By 2012 the company was advertising a series of projects in the planning stage designed to progress Pelamis towards full commercial viability:

- *Aegir*, a 10 MW wave farm off the southwest of the Shetland Islands, Scotland, to be developed jointly with utility Vattenfall. This relies on installation of a submarine cable to allow electricity export to the Scottish mainland.

- *Bernera*, a 10 MW wave farm off the west coast of Lewis in the Outer Hebrides, Scotland, with up to 14 Pelamis machines located in one of the strongest wave climates in the world – the northwest Atlantic Ocean.

- *Farr Point*, a wave farm off the Sutherland coast, northern Scotland, with a potential installed capacity up to 50 MW.

- *Aguçadoura phase 2*, a joint development with two Portuguese electric utilities of the original Aguçadoura installation, for up to 26 machines with a total capacity of 20 MW.

- *West Orkney*, development by E.ON and ScottishPower Renewables of an up to 50 MW project site off the west coast of Orkney for Pelamis technology.

In the natural world *pelamis* is the scientific name of a sea creature commonly known as the *yellow-bellied sea snake*, which is widely distributed in tropical waters. As Figure 4.1 shows, the Pelamis WEC is well-named since it takes the form of a long articulated 'snake' comprising a number of tube sections connected by universal joints. Each section contains its own power conversion equipment. Pelamis is a WEC of the attenuator type, also referred to as a *line converter*, and is kept headed into incoming waves by its front mooring connection and long thin shape – rather as a weather vane faces into the wind. It presents a small frontal cross-section to the waves and absorbs energy continuously along its length, making

Figure 4.1 A Pelamis P2 on test at EMEC (Pelamis Wave Power Ltd).

its capture width much greater than its physical width. It is particularly suited to extracting energy from long-distance ocean swells, and may be thought of as a sophisticated development of the contouring raft invented in the 1970s (see Section 1.3). As waves pass by, its movements about two axes are resisted by hydraulic rams (cylinders) that pump high-pressure oil through hydraulic motors coupled to electrical generators.

The overall design philosophy may be summarised by a few key points [3]. Pelamis is:

- Designed to be installed offshore in a range of water depths and sea bed conditions, giving flexibility over site selection.

- Constructed and commissioned off-site in safe conditions on land or in a sheltered dock.

- Rapidly attachable and detachable from its mooring and electrical connection.

- Based on established technology using tried and tested components throughout, avoiding the risks of incorporating 'prototypes within prototypes'.

Photographs of Pelamis machines at sea tend to mask their impressive physical dimensions. The P2 is 180 m long and 4 m in diameter, comprising five tube sections with a total weight of 1350 t, mostly sand ballast. A better appreciation of physical size is given by Figure 4.2 showing a machine nearing completion at the company's works in Leith, Edinburgh. Figure 4.3 shows the internal layout of one of the sections, complete with hydraulic rams, accumulators and power conversion equipment.

We have previously discussed the need for all WECs to include some sort of *reaction* system or device, obeying Newton's third law of motion that 'to every action there is an equal and opposite reaction'. A source of reaction is necessary not only for support, but also for power take-off, and in some WECs it constitutes a large part of the overall weight and cost. One important advantage of Pelamis is that the forces generated in one section are taken up by its neighbour(s) at the joints, so all reaction is inherent in the machine itself. There is no need to use the sea bed or additional massive components, either as part of the device or floating nearby.

You might expect the motion caused by waves acting on a tube section to be fairly simple. But the articulation between adjacent sections is actually very subtle and allows movements about two independent axes arranged at 90°, each with its own hydraulic ram. Furthermore, these axes are offset by some 25–30° from the horizontal and vertical. The axis allowing motion in the more vertical direction is referred to as the *heave axis*; and that allowing motion in the more horizontal direction as the *sway axis*. Combined with

Figure 4.2 A Pelamis P2 nearing completion (Pelamis Wave Power Ltd).

Figure 4.3 Internal layout of a Pelamis tube section (Pelamis Wave Power Ltd).

the buoyancy of each section, they permit complex movements in pitch, yaw and heave.

This flexibility of movement is used to great advantage for controlling power capture in a wide range of sea conditions. Back in Section 2.2.3.2 we introduced the concepts of tuning and resonance, explaining how a WEC's efficiency may be enhanced by matching its natural frequency to that of the incoming waves. In general the natural frequency of a system depends on its mass and stiffness; the lighter and stiffer, the higher the natural frequency and vice versa. Pelamis has inherent mass and hydrostatic stiffness (buoyancy) that produce a natural 'bobbing' frequency several times greater than the wave frequencies found in long-distance swells, so its two-axis system is used in a highly creative way to reduce the effective stiffness and achieve the desired degree of resonance [3]. This is done by varying the amount of restraint applied to the heave and sway axes. For example, if the restraint is equal on the two axes the machine behaves like an articulated raft and moves vertically. However, if a much weaker restraint is applied to the sway axis, the machine tends to respond with predominantly sideways or 'snaking' movements (see Figure 4.4); and since the hydrostatic stiffness about the sway axis is far lower than about the heave axis, the natural frequency is reduced. The elegance of the

Figure 4.4 Some 'snaking' movement is visible in this photo of Pelamis at sea (Pelamis Wave Power Ltd).

Pelamis articulation and control systems allows resonance to be achieved in variable sea states by adjusting the restraint applied to the two axes.

However, as we have pointed out in Section 2.2.3.2, resonance can be risky. Very large waves at resonant frequency may produce damaging forces and movements in exactly the conditions where a WEC should be limiting its response in order to survive. For this reason the Pelamis default or 'fail-safe' setting applies equal restraint to the two axes, ensuring a non-resonant response in extreme conditions. Another key to the machine's survivability lies in its fundamental hydrodynamics, which help prevent overload once rated power has been reached. As wave heights increase the tubes tend to dive below the crests and re-emerge on the far side, limiting the bending moments applied to the hydraulic rams and the power supplied to the hydraulic motors. Another noteworthy point is that Pelamis reacts to wave curvature, not height; as waves can only reach a certain steepness before breaking, this limits the range of motion that must be accommodated. Finally, the sleek shape of the device, with its small cross-section and pointed nose, is designed to minimise unwanted drag and impact (slamming) forces produced by unruly waves in high seas.

We have already mentioned that Pelamis, like other WECs of the attenuator type, has a capture width much greater than its physical width because it absorbs wave energy along its entire length. This relates to our discussion of what happens 'when waves meet WECs' in Section 2.2.3.3. Basically, a good absorber of waves must also be a good generator of waves, and the absorption process may be thought of as an interference phenomenon in which waves radiating outwards from a moving WEC partially cancel the incident waves. It follows that the detailed shape of a device's wave field has a major influence on its efficiency and capture width. The situation with a heaving point-absorber is rather simple: it generates a circular wave field, similar to that of a stone dropped in a pond. But Pelamis is quite different. It consists of a number of sections each of which would, in isolation, generate a more or less circular pattern. However, the sections are not independent – they are linked together by an articulation system that constrains them while allowing considerable individual freedom of movement. If the relative phases of their motion are carefully selected and controlled, the composite wave field can be *focussed* in a particular direction rather then spread omnidirectionally. This feature can be used to optimise energy absorption in normal sea conditions, giving a device with a length of 180 m a capture width some three times that of a simple point-absorber [3]; conversely it can be used to limit capture in extreme conditions. Doubling the length to 360 m would give a theoretical capture width five times as great. Overall, the generation of a complex wave field

Figure 4.5 Pelamis power conversion systems under construction (Pelamis Wave Power Ltd).

by a set of articulated sections helps explain why Pelamis can gather energy from its complete length, rather than just the frontal area.

Electrical connection to shore is the only viable option for a WEC located many kilometres offshore in deep water. The power take-off system of Pelamis converts the pulsating nature of wave energy into a steady electrical output suitable for feeding into a large grid. Each section of the device houses its own power conversion system (see Figure 4.5). The basic chain of events is similar to that described in Section 3.2 and illustrated in Figure 3.1. Hydraulic fluid is released from high-pressure accumulators into hydraulic motors driving 3-phase asynchronous (induction) generators rated at 125 kW each. Redundancy is included in the hydraulic system to ensure fault tolerance. Each module has a generation capacity of 250 kW and heat exchangers allow 'dumping' of unwanted electrical power in the event of cable damage or grid failure. The generated electricity is fed from the power conversion systems along the length of the machine to the nose, where it is transformed up to 6.6 kV before transmission to shore via umbilical and submarine cables.

4.2.2 Oyster

The *Oyster* oscillating wave surge converter has been researched and developed by Edinburgh-based company Aquamarine Power [5] since

2005. Essentially a large pump activated by near-shore waves, it takes the form of a hinged flap attached to a base frame piled into the sea bed. The buoyant flap, which is largely but not completely submerged, pitches backwards and forwards with the waves and activates two hydraulic cylinders. The pistons push high-pressure fresh water to the shore through a go-and-return circuit, driving a conventional water turbine and electrical generator. Oyster is typically installed about 500 m to 1 km from the shoreline in water depths of 10–15 m. The overall scheme for a multi-device wave farm is illustrated in Figure 4.6.

Oyster taps the energy in surging waves, but horizontal motion is hard to achieve in a practical WEC so pitching is used instead. In effect the hinged flap converts the surge forces exerted by waves into a *wave torque* which is transferred to the hydraulic cylinders. Not only does this simplify the mechanical design but, as we shall see, it greatly enhances the device's survivability.

The first large-scale version of the device, Oyster 1, is shown in Figure 4.7. Rated at 315 kW, it was installed at EMEC's Billia Croo wave test site in Orkney in 2009 and began delivering power to the National Grid. More than 6000 offshore operating hours in harsh Atlantic waters were achieved over the following two winters and the experience gained encouraged the company to embark on their larger second-generation device, the Oyster 800.

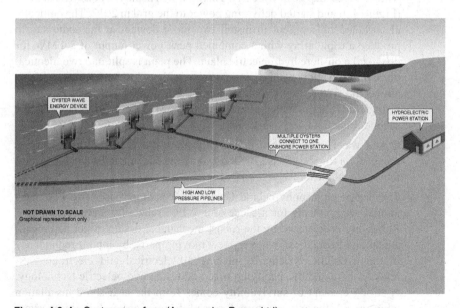

Figure 4.6 An Oyster wave farm (Aquamarine Power Ltd).

Figure 4.7 The Oyster1 device, rated at 315 kW (Aquamarine Power Ltd).

Figure 4.8 shows an Oyster 800 nearing completion. The 800 kW device, with a hinged-flap 26 m wide by 12 m high, was deployed at EMEC in 2011 (Figure 4.9) and started delivering power to the grid in 2012. The company also gained full consent to install two more devices at the same site to produce a triple array with a combined peak power output of 2.4 MW for driving the onshore hydroelectric plant. The plant is split into two identical drive trains to allow planned maintenance without losing generation. It also allows the system to generate efficiently in low-wave conditions by using a single drive train, which reduces fixed losses. In more vigorous seas the second drive train is also activated.

Oyster wave farms with installed capacities above 10 MW are firmly in prospect for the west coasts of Orkney and the Isle of Lewis, and for several international locations.

The design philosophy behind Oyster devices emphasises two key operational requirements for a successful WEC: reliability and survivability. Reliability is enhanced by the robust mechanical design of the flap mechanism and hydraulics, with very few moving parts; by inbuilt redundancy in the hydraulic system; by placing all electrical and electronic systems onshore and by specifying tried and tested hydroelectric technology. A strong element of survivability comes from the near-shore installation which avoids the extreme storm forces of the open ocean. Just as important,

Figure 4.8 The Oyster 800 device, rated at 800 kW (Aquamarine Power Ltd).

the hinged-flap design has an inherent safety feature due to its pitching motion. Waves rotate the flap downwards towards the sea bed before it bobs back up again. As wave heights increase they tend to push the flap down further, causing more and more water to spill over the top of the device and limiting the amount of power intercepted. This feature reduces the need for complex control systems, and avoids shut-down of the device in storm conditions.

Oyster's near-shore wave climate has some special features which bear importantly on its design:

- Sea bed friction and wave breaking close to shore tend to dampen extreme conditions. Wave heights at the 12 m depth contour during storms are typically only 50% of those far offshore, improving survivability.

- An average annual offshore wave power resource of, say, $70\,\mathrm{kW\,m^{-1}}$ (see Figure 1.7) includes infrequent but extremely powerful storm events that cannot be tapped by a WEC – it must focus on survival. Design should aim at maximising a device's annual capacity factor (see Section 1.4), not trying to capture extremes. Although the annual near-shore resource is considerably smaller than that offshore, a greater proportion of it is *exploitable* for electricity production.

Figure 4.9 Oyster 800 installed at EMEC's wave test site (Aquamarine Power Ltd).

- Wave energy tends to be much more consistent in direction near-shore than offshore. Directional spread reduces as waves enter shallow water due to refraction effects (see Section 2.1.4), making it unnecessary for a directional near-shore WEC to 'swing round' to face them.

An ongoing R&D programme in conjunction with Queen's University Belfast, including numerical modelling and tank testing, has led Aquamarine Power to an advanced understanding of hinged-flap WECs [6, 7]. The overall design aim for Oyster 800 is to achieve a balance between technical performance, capital cost, the cost of energy and availability. Key technical factors include:

- *Flap design.* A wide flap is generally more efficient at capturing the wave power across its width than a narrow one, although it tends to reduce performance in short-crested sea states, or where wave crests are not parallel to the flap, because forces generated along the flap width may not all be in phase. Flap efficiency may be improved by adding wide rounded ends, and it is important to prevent water leakage over, under or through the flap in light and moderate seas. The hinge should be located as close as possible to the sea bed to maximise wave torque. Oyster 800's flap dimensions of 26 m × 12 m are sufficient to power a device rated at 800 kW.

- *Water depth.* For reasons already noted it is desirable to operate in fairly shallow water. Also, the increase in horizontal water particle motion due to shoaling can increase surge forces by up to 50%, although the benefits reduce in water depths less than about 10 m as waves start to break.

- *Tuning.* It is generally considered desirable to tune WECs to the frequency of incident waves (see Section 2.2.3.2). Near-shore devices with bottom-hinged flaps tend to have a large mass (moment of inertia) relative to stiffness (buoyancy), mainly because the effective mass is greatly increased by the volume of water that moves with the flap. This produces a much lower natural frequency than that of commonly occurring waves, implying that hinged-flap devices normally operate 'de-tuned'. However, research has demonstrated that the average wave forces and wave torque generated over a range of sea states largely determine Oyster's performance and that maximising them is a better strategy than conventional tuning, which is more relevant to offshore devices operating in the near-linear waves of long-distance swells.

- *Energy absorption.* Effective absorption of energy by a WEC depends on controlling the amount of damping (see Section 2.2.3.2). In Oyster's case the damping is controlled by varying the working pressure in the hydraulic cylinders. This offers the chance of optimising power output from the device on a continuous basis by matching damping to current wave conditions.

The energy that Oyster extracts from the waves is delivered to shore using a closed loop hydraulic system, and the conversion from hydraulic to electrical energy occurs within an onshore generating station. This avoids any corrosion, shock and bump issues and allows complex mechanical, electrical and electronic systems to be easily maintained.

The hydraulic energy is converted to rotational energy through a Pelton turbine. Discharge from the turbine is stored in a header tank, to allow any entrained air to escape from the fluid before returning to the Oyster. The drive train, shown in Figure 4.10, connects the Pelton wheel to a flywheel and generator. The motive torque can accelerate the drive train and store rotational energy in the flywheel, allowing smoothing of electrical power delivery, or it can be used directly to generate power.

The power generation system is based on a robust squirrel-cage induction generator (see Section 3.4.3.1) and a full-scale, low-voltage (690 V), electronic power converter which controls the power extracted from the drive train. The system is arranged to use a wide operating speed range of 1500–3600 rpm to maintain high Pelton turbine efficiency across all sea

Figure 4.10 The Oyster 800 onshore drive train (Aquamarine Power Ltd).

Figure 4.11 Oyster 800 arrives at Orkney (Aquamarine Power Ltd).

Figure 4.12 A dramatic scene at Billia Croo, EMEC's wave test site (Aquamarine Power Ltd).

states. The system pressure is maintained by active control of the inlet valves to the Pelton wheel. In smaller sea states a single drive train operates at around 1800 rpm and system pressure is regulated to about 60–80 bar; in higher sea states two drive trains operate with the pressure regulated to about 100–120 bar. The generator and power converter both use process water from the header tank for cooling, keeping the generation building sealed from the salty environment and enhancing the overall reliability of the system.

Figures 4.11 and 4.12 show Oyster 800 being towed towards the wave test site, and in its dramatic test location.

4.2.3 Limpet and Mutriku

Oscillating water columns (OWCs) have a long history and are some of the best researched of all wave energy devices [2, 8]. Two pioneering modern examples have been developed by Voith Hydro Wavegen [9], a company based in Inverness, Scotland. The first, installed on the Hebridean island of Islay in 2000 and known as *Limpet*, was the first commercial-scale wave energy plant in the world to be grid-connected. The second is a much larger installation commissioned in 2011 at Mutriku, on the northern coast of Spain between San Sebastian and Bilbao.

We have already covered basic OWC principles in Section 2.2.2 (see also Figure 2.15). A hollow structure, open to the sea below the water line, encloses a column of water topped by a column of air. Incoming waves cause oscillations of the water column which compress and decompress the air, driving a turbine. Converting the hydraulic force of waves into air pressure neatly overcomes one of the main challenges of wave energy conversion – marrying *slow-speed* waves to a conventional *high-speed* electrical generator. The unique form of construction and power take-off makes onshore OWCs substantially different, in theory and practice, from other types of WEC.

Limpet's position on the wild west coast of Islay, built slightly inland from the natural shoreline in a man-made recess, is shown in Figure 4.13 (for the location of Islay, see Figure 1.11). The hollow structure admits waves below an *entry lip* onto an inclined ramp, alternately compressing and decompressing the air column above and driving air forwards and backwards through the turbine duct – see Figure 4.14a. The roof is strengthened by internal supports dividing the water column into three equal chambers 6 m wide, and the front wall is robust enough to withstand wave 'slamming' in storm conditions. Heavy civil engineering work was required to excavate the rock and build the reinforced concrete structure [10].

Two essential conditions for a successful WEC are well served by Limpet's location and technology:

- *Reliability.* There are no moving parts in contact with the waves. The turbine and all electrical and electronic components are onshore. The turbine-generator technology is well tried and tested.

- *Survivability.* The structure is extremely strong – shortly after commissioning Limpet experienced a 'once-in-50-year' storm.

One of *Limpet's* most important features is the *Wells turbine* that drives the electrical generator. We have covered the operating principle of this type of turbine in Section 3.2 and, for convenience, show a typical rotor again in Figure 4.14b. A number of blades with symmetrical airfoils are mounted so their chord lies in the plane of rotation. The blades generate similar lift forces with the air flow in either direction, making the turbine 'self-rectifying' – it speeds up and slows down during each air pressure cycle, but never reverses. It can, therefore, be used to drive a conventional variable-speed generator.

Figure 4.13 Limpet clings to the rocks of Islay (Voith Hydro Wavegen Ltd).

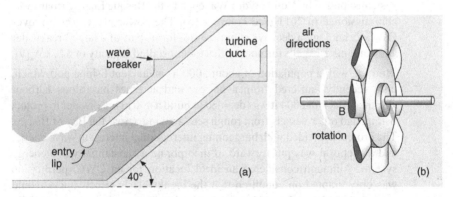

Figure 4.14 (a) Limpet's oscillating water column and (b) a Wells turbine rotor.

Various turbines with ratings between 500 kW and 18.5 kW have been installed and tested since Limpet was commissioned in 2000, reflecting its importance for wave energy research as well as supplying electricity to the local grid. Much theoretical work on OWCs, including numerical modelling, has been done over the years with the aim of understanding and predicting performance [2]. However, it is very difficult to relate the complex hydrodynamic and pneumatic effects to the characteristics of a variable-speed air turbine, especially when nonlinear waves in random seas are considered.

Back in Section 2.2.3.3 we discussed the reflection and absorption of energy that occur when waves meet a WEC. For effective energy absorption it is necessary for a device to 'generate its own waves' with the correct frequency and phase; it should neither dissipate energy in turbulence and friction, nor reflect it back again. This idea has big implications for shoreline devices such as Limpet which present an obvious 'barrier' to incoming waves. The working area of the barrier cannot be rigid – we do not wish to see large waves reflected back out to sea – rather, it should resemble an oscillating flap or piston. Yet merely to state this shows how complicated the situation is when the working area consists of a body of water pulsating beneath an entry lip and meeting highly variable air pressure on the inside. Figure 4.14a may look simple, but it hides a multitude of complexities! For this reason performance data collected from Limpet over more than a decade, with more than 60 000 hours of grid connection, have proved invaluable to wave energy R&D.

We now move to Mutriku on the northern coast of Spain, where an OWC system supplied by Voith Hydro Wavegen to the Basque Energy Board was commissioned in 2011 (see Figure 4.15). The power plant, which moves OWC technology a decisive step forward, forms part of a large breakwater and contains 16 Wells turbines with a total installed capacity of 300 kW [9].

Mutriku, with a population of about 5000, is an ancient fishing port which, until recently, suffered from a narrow and at times hazardous harbour entrance. Back in 2004 it was decided to build a 400 m breakwater to protect fishing and other vessels from rough seas arriving across the Bay of Biscay. The project coincided with burgeoning international interest in wave energy and a proposal was put forward to incorporate a substantial wave energy system. Although considered an ideal location for an OWC, permission was only granted on condition that the breakwater's effectiveness would not be jeopardised. This made the size and placing of the OWC especially important because of concern that waves reflected back from the structure in storm conditions might have serious effects on navigation outside the harbour. It was agreed that a 100 m length of the breakwater, well away from the main navigation channel, could be modified. Unlike the rest of the

Figure 4.15 The power plant at Mutriku on the northern coast of Spain (Voith Hydro Wavegen Ltd).

sloped breakwater it would present a vertical, slightly curved, frontal area to the incoming waves.

At this stage much effort was put into designing the OWC structure and estimating its potential energy yield [11]. The design centred around a set of 16 OWC chambers (as opposed to Limpet's 3) with openings below sea level at all times, topped by 16 vertically-mounted Wells turbines rated at 18.5 kW each.

The performance of a WEC depends crucially on the wave climate, including power levels and spectral composition, in which it operates (see Section 2.1). Mutriku's near-shore power density averages about 18 kW m^{-1} in the winter months, reducing to 5 kW m^{-1} in summer, and is strongly directional from the WNW. Researchers were able to describe the annual variations by a set of 14 representative sea states covering 63% of the year (the other 37% has waves that are too small to be useful); the most powerful sea state has a significant wave height of 3.3 m and period of 12.5 s; the least powerful, a significant wave height of 0.88 m and period of 5.5 s. The information was used in a series of wave tank tests on 1/40th scale models which gave a predicted annual pneumatic power capture of 12.2 kW for each OWC chamber and, assuming 90% availability, a total of 174.5 kW for the whole plant [11]. Multiplying this by the number of hours in a year (8760) gave an estimate of 1530 MWh for the annual pneumatic energy capture. Finally, full-scale turbine-generator tests carried out at the Limpet plant on Islay indicated a conversion efficiency close to 40% from pneumatic to electrical power, indicating that Mutriku would be able to deliver 600 MWh of renewable electricity in a full year.

The completed plant has 16 vertically-mounted 18.5 kW Wells turbines, each with twin five-bladed rotors separated by an air-cooled generator and fitted with a noise attenuator (Figure 4.16). Turbine control is based on instantaneous chamber pressure and optimises rotation speed to produce maximum power output. The variable-speed induction generators, rated at 450 V, are connected in two groups of 8. Their variable-voltage, variable-frequency, outputs are rectified and inverted by power converters and transformed up to 13.2 KV for feeding into the local 50 Hz grid.

Figure 4.16 The Mutriku turbine gallery (Voith Hydro Wavegen Ltd).

As completed, the Mutriku plant produced 100 MWh of grid electricity during the commissioning phase and Voith Hydro Wavegen supplied the client with commercial guarantees for performance and reliability. By the end of 2012 the company was estimating annual electricity production equivalent to the consumption of 250 households [9]. The project illustrates the benefits of integrating wave energy technology into local infrastructure – the breakwater was going to be built anyway, so to add a WEC system made economic as well as environmental sense. The inhabitants of Mutriku now have a much safer harbour, plus the promise of increased revenue from tourism as the fame of their OWC spreads around the world.

4.2.4 Wave Dragon

The Danish device known as *Wave Dragon* [12] is one of the best known, and most developed, WECs based on the overtopping principle. Waves are captured, directed by a specially-shaped ramp into a reservoir above mean sea level (MSL), and returned to the sea via hydroelectric turbines (see Figure 4.17). In effect power take-off is achieved by converting the surging energy of waves into the potential energy of a head of water. The effective capture width is greatly increased by wave reflector wings attached to the sides of the reservoir, clearly visible in the large prototype device shown in Figure 4.18.

Overtopping WECs may, in principle, be built on shorelines or, as intended for a full-scale Wave Dragon, moored out at sea in water depths of 25 m or more. Offshore deployment generally offers far better wave resources and a greater range of suitable locations for large devices but, as with other floating WECs, survivability in storm conditions tends to become a major part of the design challenge.

Unlike other offshore WECs, Wave Dragon is designed to remain stationary and stable rather than heave or pitch with the waves. However, a key feature is its ability to alter the floating level of the platform, and hence the height of the reservoir above MSL, by blowing air into, or venting air from, open chambers underneath the device. The overall aim

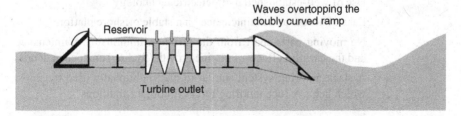

Figure 4.17 Wave Dragon works on the overtopping principle (Wave Dragon ApS).

Figure 4.18 A prototype of *Wave Dragon* showing the two wave reflector wings (Wave Dragon ApS).

is to maximise the amount of overtopping and flow through the turbines in line with prevailing wave conditions, without overfilling the reservoir and losing water by spillage. In gentle seas the device is set low in the water to encourage overtopping by small waves; in rough conditions it is raised to take advantage of increased wave heights and a better pressure head for driving the turbines. The ability to adjust the height of the reservoir represents a major advance over a fixed shoreline device such as a tapering channel (see Section 1.3.1 and Figure 1.15) and offers great potential for optimal control to maximise the device's performance.

Other key features of Wave Dragon include:

- A simple and robust design which captures wave energy directly.
- Use of well-established hydroelectric technology.
- Ease of access and maintenance on a stable offshore platform.
- No moving parts apart from the turbines, minimising maintenance and the harmful effects of marine growth (fouling) and ocean debris.
- A slack mooring system similar to that traditionally used by ships, which helps reduce mooring forces in rough conditions.

- Good survivability because extreme waves simply ride over the top of the device.

Detailed design work on Wave Dragon began in 1986. During the 1990s extensive studies were carried out on major aspects of a projected multi-megawatt device, including basic geometry and structural design, reflector wing efficiency, turbine configuration, survivability and economics [13]. In 1998–1999 a 1/50th scale model was tested at Aalborg University in Denmark and the results were sufficiently encouraging to set up a six-nation European project to design and build a large (but not yet full-scale) prototype. In 2003 the device, at 1/4.5th scale and weighing 237 tonnes, was deployed in the *Nissum Bredning*, a large inland sea in Jutland with an area of some 250 km². This is the device shown in Figure 4.18. The degree of scaling, which gave a total width across the reflector wings of 58 m, was chosen to suit Nissum Bredning's relatively gentle wave climate. It is estimated that a full-scale equivalent made in concrete rather than steel, weighing 33 000 tons and 300 m wide, would generate up to 7 MW if deployed offshore in a typical North Atlantic climate with an average power density of 36 kW m^{-1}.

The Nissum Bredning prototype allowed many key components and sub-systems to be assessed realistically before committing funds to a full-scale version deployed in the open sea. The device yielded extremely valuable data over more than 20 000 hours of operating experience, including grid connection, which verified the predicted overtopping and wave absorption performance. Following minor design modifications Wave Dragon may reasonably claim to be the best researched and developed of all offshore overtopping WECs [13]. Although the global financial crisis of 2008 affected the company's plans to test a full-scale multi-megawatt device off the coast of Pembrokeshire in Wales, in 2011 it announced an alternative 1.5 MW version for deployment at the Danish Wave Energy Center (danWEC) at Hanstholm on Jutland's North Sea coast.

We may summarise the technical design and performance of Wave Dragon under several headings [13]:

- Overtopping and ramp design.
- Wave reflector wings.
- Turbines.
- Electricity generation.
- Control and performance optimisation.

207

Overtopping of waves into a reservoir – especially in random, real-life seas – is a highly nonlinear process which does not easily lend itself to theoretical analysis. Successive waves vary in height in a random manner; some may overtop the ramp vigorously, others not at all. The best approach is to obtain empirical data from scale models in wave tanks, describing the actual overtopping flow that occurs in a given wave climate (significant wave height and period) in statistical terms. Using this information, the shape of the ramp and the height of the reservoir above MSL may be adjusted to maximise overtopping and turbine power output in a variety of sea states. This approach led to a patented double-curved design for Wave Dragon's ramp (see Figure 4.19) which has proved highly efficient at capturing the incident wave energy.

The long, slender, wave reflector wings on Wave Dragon are very cost-effective. By increasing the height of waves approaching the ramp they greatly improve the amount of overtopping and energy capture. Their buoyant design stabilises the main platform, reduces water spillage, helps

Figure 4.19 A close-up view of the doubly-curved ramp on the 1/4.5th scale device (Wave Dragon ApS).

keep the device aligned with the oncoming waves and reduces peak mooring forces. Extensive computer modelling was used to maximise the amount of wave energy directed towards the ramp by the reflector wings in a variety of sea states. It turns out that relatively small high-frequency waves which occupy much of the year are the most enhanced, mainly because their energy, being concentrated near the sea surface, is captured by the wings rather than passing below them.

The turbines on Wave Dragon must operate efficiently over an unusually wide range of pressure heads – typically 1 m to 4 m in a full-scale device – as the water level in the reservoir fluctuates. The type chosen is the vertical-axis *Kaplan turbine*, used extensively in low head hydroelectric plants for more than 80 years, but simplified to minimise the number of bearings and oil seals that must function reliably in the hostile salt-water environment. Rather than install a single large turbine, there are important advantages in using a number of smaller machines to share the available flow:

- Maintenance may be carried out on individual turbines without disrupting electricity production.
- Smaller, lighter, machines are easier to hoist and transport.
- Small turbines rotate faster than large ones and may be directly coupled to electric generators without the need for gearboxes.
- At low flow rates, overall device efficiency is improved by reducing the number of turbines in operation.

The number of turbines installed varies according to the size of Wave Dragon, from 7 units rated at 2.3 kW each supplied from a 55 m^3 reservoir in the Nissum Bredning prototype; to 16–20 units rated at 400 kW each that would be required for a future 7 MW offshore device with a reservoir capacity of 8000 m^3. Individual turbines may be turned off, or have their speeds continually adapted to the available head in the reservoir, maintaining efficiencies close to maximum value.

Electricity production is by variable-speed, permanent-magnet, synchronous generators coupled directly to the turbines. Since turbine speeds vary widely as the water level in the reservoir fluctuates, so does the frequency of the generated electricity. The output of each generator is, therefore, processed by a full-scale electronic power converter (as discussed for tidal stream turbines in Section 3.4.2) to bring it to the desired grid frequency, and transformed up to a suitable voltage for transmission.

The final topic in the earlier list – control and performance optimisation – is arguably the most significant of all. As already noted, air chambers underneath the reservoir allow its height above MSL to be adjusted in line with the prevailing wave conditions. Back in Section 2.2.3.2 we discussed the *tuning* of WECs to optimise performance, pointing out that some sophisticated devices incorporate both *slow* and *fast* tuning: the former adapts to relatively slow variations in sea state; the latter to individual waves or small groups of waves. Raising and lowering Wave Dragon's reservoir over time scales up to a few hours may be considered a form of slow tuning, similar in principle, although different in practice, from that applied to heaving and pitching devices. Adjusting the water level in the reservoir, achieved by rapid control of flow through the turbines, is effectively a form of fast tuning. For example, by lowering the water level when some large waves are expected, spillage is minimised and their energy is captured. Intelligent control of reservoir height and water level using programmable logic controllers (PLCs) and an internet-connected system control and data acquisition (SCADA) system helps optimise Wave Dragon's overall performance and enhance survivability in extreme conditions [13], for little or no increase in capital cost.

More than 20 000 hours of operating experience in Nissum Bredning with the 1/4.5th scale device confirmed the survivability of Wave Dragon's general design. The energy of extreme waves tends to pass over or under

Figure 4.20 Wave Dragon survives a rough sea (Wave Dragon ApS).

the device, avoiding the 'end-stop' problems that beset other types of WEC. A few incidents did occur, especially when a 100-year storm hit the Jutland coast of Denmark in 2005, but no serious damage was inflicted and valuable lessons were learned about the need to incorporate redundancy in key components of such a complex engineering system (Figure 4.20).

4.2.5 PowerBuoy®

The *PowerBuoy*® family of point-absorber WECs has been under development by the American company Ocean Power Technologies Inc. [14] since 1994. The devices are essentially floating buoys, anchored to the sea bed, which generate electricity as they heave up and down in the waves. Relatively low-power versions rated up to 40 kW are primarily aimed at the market for 'autonomous' deep-ocean applications including maritime security, oil and gas operations and oceanographic research. In recent years the company has progressed towards larger, higher-power versions aimed at the utility market and by 2012 it had manufactured and tested a 150 kW grid-connected device, with a 500 kW version known as *PowerTower*® at an advanced stage of development. Major projects completed or in progress include:

- A 40 kW version delivered to the US Navy and deployed off Hawaii.
- A prototype 150 kW device deployed and tested off the north-east coast of Scotland.
- Successful deployment in Spain of an underwater substation pod to interconnect up to 10 devices, with transmission of power and data to shore by a single submarine cable.
- A 150 kW device to be deployed at the *Wave Hub* test site located off the coast of Cornwall, England.
- A wave farm of ten 150 kW devices to supply electricity to a coastal community at Reedsport, Oregon, USA.
- Supply of devices for a planned 19 MW wave energy plant off the coast of Victoria, Australia.

We focus here on the 150 kW device referred to as the *PB150* and designed to operate in water depths above 55 m, which was tested in the open North Sea some 33 nautical miles off the north-east coast of Scotland in late 2011. The photo in Figure 4.21 shows the device lying horizontally on the quayside prior to being towed offshore, and the illustration in Figure 4.22 indicates its principal dimensions. Note that, when deployed and floating vertically, it extends some 35 m below MSL, with a float 11 m in diameter.

Figure 4.21 A 150 kW PowerBuoy® ready for towing out to sea (Ocean Power Technologies Inc).

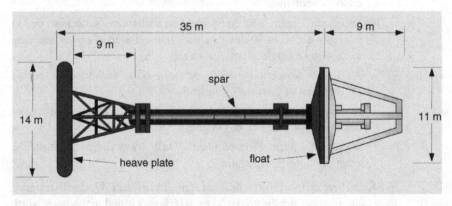

Figure 4.22 Principal dimensions of the 150 kW device.

The device is designed for significant wave heights between 1 and 6 m in normal operation, and during six months of sea trials was reported to average 45 kW of electrical power output in significant wave heights of 2 m, with peak levels in excess of 400 kW.

Back in Section 2.2.3.3 we noted that wave-generated forces in a WEC must always be opposed by equal reaction forces, either supplied from within the device or external to it. In the case of the PB150, which is moored but not rigidly attached to the sea bed, reaction forces are provided by a large *heave plate* (see Figure 4.22), placed sufficiently low in the water to avoid being significantly affected by the waves. Relative motion between heave plate and float powers the device. Another dimension that bears importantly on device performance is the float diameter. Our discussion of floating bodies in Section 2.2.3.1 explained why the physical size of a device in the wave direction can have a big effect on the type and amount of motion. In the case of a heaving device, movement may become quite restricted if the float diameter exceeds about one fifth of a wavelength. This implies that the PB150 is best suited to the longer wavelengths – say above 60 m – in random seas. We should also remember that the capture width is greater than the physical width of the float because the device interacts with the entire surrounding wave field (see Section 2.2.3.3). It is also important to bear in mind an advantage of point absorbers over most other types of WEC: axial symmetry makes them omnidirectional and able to collect wave energy equally from any direction.

Back in Section 2.2.3.2 we introduced the topic of tuning, explaining that power capture is generally enhanced when the natural frequency of a WEC coincides with the wave frequency. The concept is especially applicable to a point absorber which has a well-defined natural 'bobbing' frequency due to the float's inherent mass and stiffness (buoyancy). When once the wave climate in which it is to operate is known, the natural frequency may, in principle, be set to match the waves that are expected to provide the greatest amount of energy over a full year. Of course real-life seas are generally random with a large range of frequencies (and periods), so 'fixed tuning' can be only partially successful. A more sophisticated approach incorporates variable tuning in the design, allowing the device to adapt to changing sea states and maximise the annual energy capture. Such adaptation can be either 'slow', responding to changes in a time scale of hours; or 'fast', attempting to follow groups of waves or even individual waves over much shorter time scales. One important caveat about tuning is that it increases the amplitude of oscillations, which may become substantially greater than the wave height, leading to vulnerability in high seas unless the system is adequately damped or provided with effective 'end stops'. Much of the design effort for the PB150, and its larger cousin the 500 kW *PowerTower*®, addresses these issues with effective high-tech electronic control solutions, including fast 'wave-by-wave' tuning. This gives efficient energy capture in a wide range of sea states, coupled with a high degree of survivability in severe storms and even hurricanes.

Power take-off in WECs takes a variety of forms, but the simple heave motion of a large point absorber device is well suited to a mechanical system driving AC generators, either conventional or linear (see Section 3.4.4). In the PB150 the AC output is rectified and inverted by a full-scale power converter, producing fully grid-compliant electricity at 575 V, 50 Hz or 600 V, 60 Hz, with a power factor between unity and ±0.9. The generator is rated at 150 kW maximum daily average, with short-term peaks up to 866 kW.

To summarise, present high-power versions of PowerBuoy® include the following features:

- Omnidirectional energy capture.
- Fast tuning by advanced electronic control.
- Direct-drive power take-off.
- Grid-compliant electricity transmitted to shore.
- Demonstrated survivability in extreme storms and hurricanes.

Figure 4.23 shows a PB150 being towed out to sea by a standard tug and Figure 4.24 shows it on location and ready to generate.

Figure 4.23 Towing the PB150 out to sea with a standard tug (Ocean Power Technologies Inc).

Figure 4.24 Ready to start generating: the 150 kW PowerBuoy® (Ocean Power Technologies Inc).

4.2.6 Penguin

The WEC known as *Penguin* has been developed since 2008 by the company Wello Oy [15] based in Espoo, Finland. Shaped as a floating asymmetric vessel, it houses an eccentric rotating mass and electrical generator co-mounted on a vertical shaft. As waves pass by, movement of the vessel causes the mass to rotate, generating electricity. A sophisticated control system is designed to maximise power take-off in a wide range of sea states.

Scale models were used to validate the Penguin concept over the period 2008–2011, starting with a 1/18th scale device used in a series of wave tank tests. This was followed by a one-eighth scale model deployed in sea conditions where it survived the equivalent of a 100-year storm on three separate occasions. The results encouraged the company to design and build a full-scale 1 MW prototype which was launched at the Riga shipyard in Latvia in 2011 and towed 2000 km across the Baltic and North Seas to EMEC in Orkney.

Figure 4.25 shows Penguin moored in Orkney in June 2012, ballasted and ready for towing to EMEC's Billia Croo wave test site, and Figure 4.26 shows it moving out to sea. Key features include:

- Weight 1600 t including 1300 t of ballast; length 30 m; height 9 m of which only 2 m is visible above the waterline.

Figure 4.25 The fully-ballasted Penguin at Lyness wharf in Orkney (Wello Oy).

Figure 4.26 Penguin goes to sea (Wello Oy).

- Construction using traditional steel fabrication and shipbuilding techniques available in shipyards worldwide.
- Easily-scalable manufacturing.
- All operational parts within the hull's protective cover.
- Simple power take-off involving only one moving entity (coupled mass and generator).
- Electrical and power conditioning equipment similar to that developed over many years for the wind turbine industry.
- Overall economy due to specification of standard components.

Initial test results were consistent with data obtained from the scale models, with the added bonus that efficient rotation of the eccentric mass occurred in smaller waves than originally anticipated. By the autumn of 2012 the device and its three-cable slack mooring system had survived their first major storms in waves up to 12 m high, and it was decided to complete commissioning of the power-plant in gentler conditions at the quayside. By the end of the year the company reported highly satisfactory progress, with a full programme of work scheduled for 2013 and beyond.

Figure 4.27 Internal layout of the full-scale prototype Penguin, showing the eccentric mass (rotator) and synchronous generator (Wello Oy).

The internal layout of the full-scale prototype is illustrated in Figure 4.27. The eccentric mass or *rotator* (coloured red) turns continuously in one direction, seeking the lowest point under the influence of gravity and the main shaft's inclination from the vertical. Its mass of 120 t may, in principle, be increased to a maximum of 200 t. The rotator completes one revolution for each passing wave, giving a typical speed range of 5–15 rpm. Since rotation speed is tied to wave frequency, the control system optimises power take-off by maximising the torque on the main shaft (mechanical power being proportional to both speed and torque). This is done by introducing a *phase lag* – equivalent to an angular displacement – between the rotator's instantaneous position and the shaft's angle of inclination. In small waves the phase lag is close to 0°, but in large waves it increases towards 90°. The control strategy is designed to give the maximum possible torque – and hence power output – over a wide range of sea states.

The rotator, which also acts as a flywheel, is directly coupled to the generator (shown yellow in Figure 4.27). There is no gear box. The 1 MW generator is of the permanent-magnet synchronous type with 40 poles. The typical speed range of 5–15 rpm produces a variable frequency output which, for grid compatibility, is rectified and inverted by a full-scale electronic power converter (see Section 3.4.2) and exported to shore by subsea cable at 11 kV.

Back in Chapter 2 we discussed a number of key issues that affect the performance of WECs, including the movement of floating devices in waves of various wavelengths, the amount of tuning and damping, and the concept that 'to absorb a wave is to generate a wave'. These issues are relatively easy to visualise and analyse in the case of a point absorber that simply heaves up and down in the waves. But, as Figure 4.28 suggests, *Penguin* often displays complex patterns of movement in heave, pitch and roll – indeed, its 'awkward' shape is deliberately designed to extract energy from all movement modalities. Any discussion of the above issues therefore risks oversimplification, and is further constrained by the introductory level of this book. Nevertheless, you may find the following comments helpful:

- *Device dimensions and the effects of different wavelengths.* As already noted, the prototype full-scale Penguin is about 30 m long. Assuming that, in a regular swell, it faces the oncoming waves, this dimension has a substantial effect on the amount it heaves and pitches in waves of various wavelengths (see Section 2.2.3.1). In random seas the situation becomes far more complicated.

- *Tuning and damping.* As with other floating WECs, the degree of tuning depends on mass and buoyancy (stiffness). Penguin has many degrees of freedom in heave, pitch and roll, each with its own mass (or moment of inertia) and buoyancy (or restoring moment). Damping is a necessary condition for energy capture (see Section 2.2.3.2) and may also be used to limit movement to safe levels in storm conditions. In *Penguin* the amount of damping associated with each degree of freedom is substantially determined by the basic design and shape of the vessel, but modified by the damping effect of the carefully controlled 'phase lag' between rotator and shaft inclination which is used to maximise power take-off.

| (a) | (b) |

Figure 4.28 Penguin heaves, pitches and rolls (Wello Oy).

- *To absorb a wave is to generate a wave.* The complex shape of Penguin's hull, with its many degrees of freedom, makes it hard to visualise this important principle in action. Yet to maximise energy yield, the device must generate a pattern of waves that interferes as effectively as possible with the surrounding wave field (see Section 2.2.3.3).

Even to mention the above issues suggests the difficulty of modelling Penguin's energy capture theoretically. In such cases, practical measurements on scale models in wave tanks and sea conditions are particularly relevant, especially when backed up by operational experience with a full-scale prototype in a variety of sea states. An intensive and ongoing R&D programme gives Wello Oy an unrivalled understanding of its innovative and unusual device.

Survivability also looms large in the mind of the WEC designer and Penguin is no exception. The device's survivability is enhanced by a number of design features:

- The vessel's type and shape, and its movement in extreme waves.
- A mooring system which allows the device to follow wave motion fully.

Figure 4.29 Penguin on test at EMEC's Billia Croo wave test site (Wello Oy).

- An extremely robust rotator-generator system. The rotator does not accelerate in storm conditions; rather, the long period of big waves causes it to slow down.

These features, together with the one-eighth scale model tests in storm conditions back in 2010 and the full-scale prototype's survival in 12 m waves in 2012, give Wello Oy strong grounds for optimism (Figure 4.29).

References

1. Pelamis Wave Power Ltd, www.pelamiswave.com (accessed 24 April 2013).
2. J. Cruz, ed. *Ocean Wave Energy: Current Status and Future Perspectives*, Springer: Berlin (2008).
3. R. Yemm. *Pelamis*, Chapter 7 in ref. 2 above, Springer: Berlin (2008).
4. EMEC, www.emec.org.uk (accessed 24 April 2013).
5. Aquamarine Power Ltd, www.aquamarinepower.com (accessed 24 April 2013).
6. A. Henry, L. Cameron, R. Doherty, T. Whittaker, K. Doherty. *Advances in the Design of the Oyster Wave Energy Converter*, Edinburgh: Aquamarine Power Ltd (2010).
7. L. Cameron, R. Doherty, A. Henry. *Design of the Next Generation of the Oyster Wave Energy Converter*, 3rd International Conference on Ocean Energy, Bilbao, Spain (2010).
8. Wave Energy Centre, www.pico-owc.net (accessed 24 April 2013).
9. Voith Hydro Wavegen Ltd, www.wavegen.co.uk (accessed 24 April 2013).
10. T. Heath. *Oscillating Water Column – LIMPET*, Chapter 7 in ref. 2 above, Springer: Berlin (2008).
11. Y. Torre-Enciso, I. Ortubia, L.I. López de Aguileta, M.J. Mutriku. *Wave Power Plant: From the Thinking out to the Reality*. Proceedings of the 8th European Wave and Tidal Energy Conference, Uppsala, Sweden (2009).
12. Wave Dragon, www.wavedragon.net (accessed 24 April 2013).
13. J. Tedd et al. *Wave Dragon*, sections 7.4 and 7.5.5 in ref. 2 above, Springer: Berlin (2008).
14. Ocean Power Technologies Inc, www.oceanpowertech.com (accessed 24 April 2013).
15. Wello Oy, www.wello.eu (accessed 24 April 2013).

5

5 Case studies: Tidal stream energy converters

5.1 Introductory

Tidal stream devices have seen great advances in the past decade and are presently at a stage of development where many full-scale prototypes are being deployed for demonstration and performance measurement, with a realistic prospect of commercial-scale multi-megawatt arrays in the next few years. The aim of this chapter is to describe a number of devices at the forefront of development, many of which are being, or have been, tested at the European Marine Energy Centre (EMEC) in Orkney, and relate them to the scientific principles discussed in previous chapters. We will also raise a variety of practical issues that greatly affect device developers as they build, deploy and test devices which, in many cases, approach or exceed 1000 tonnes in weight.

Tidal stream devices tap the kinetic energy of tidal flows and the majority of today's advanced designs mimic the action of large wind turbines, even though the fluid that powers them is about 800 times denser and moves more slowly. These differences affect the size of their rotors, whereas the power trains – gearboxes, generators and electronic power converters – are largely unaffected by whether it is wind or tide that drives them. It is, of course, a principal reason why we may expect tidal stream engineering to be influenced by wind engineering in the coming years. This is not to deny that alternative designs will sometimes challenge the norm, and some may succeed, but it will be hard for them to reach, let alone overtake,

Electricity from Wave and Tide: An Introduction to Marine Energy, First Edition. Paul A. Lynn.
© 2014 John Wiley & Sons, Ltd. Published 2014 by John Wiley & Sons, Ltd.

what is fast becoming a 'standard recipe': the horizontal-axis turbine. This is in stark contrast to the wave energy scene where devices take a wide variety of forms and work on a range of scientific principles – depending, to a large extent, on whether they are deployed on shore, near-shore or offshore.

When we note that tidal stream devices are tending towards a standard recipe, we are referring to their horizontal-axis rotors, power trains and nacelles. But as we shall see in the following pages, current designs differ greatly in other respects: the pitching and profiling of rotor blades; whether or not the nacelle turns twice during each tidal cycle to face the flow; methods of deploying, securing and recovering the turbine for maintenance and inspection; and designing for survivability – to mention just a few key issues. All this makes the total design package surprisingly variable, even though the underlying physics of energy capture is shared.

5.2 Case studies

5.2.1 Andritz Hydro Hammerfest

The Norwegian company Hammerfest Strøm [1], based above the Arctic Circle in the coastal town of Hammerfest, was established in 1997 and deployed its first tidal stream turbine, the *HS300*, in 2003. Installed in Kvalsund, a narrow fjord some 30 km outside Hammerfest, the 300 kW grid-connected device, illustrated in Figure 5.1, provided 17 000 h of operational experience and showed an availability of 98% during prolonged reliability testing. This encouraged the company to develop its 1 MW full-scale prototype, the *HS1000* (Figure 5.2). In 2012 ANDRITZ HYDRO, an international supplier of hydropower machinery headquartered in Graz, Austria, became the majority shareholder in the company, which is now known as ANDRITZ HYDRO Hammerfest.

The HS1000 is essentially a scaled-up version of the earlier HS300 and was first deployed at EMEC's Fall of Warness tidal test area [2] in December 2011. It is a three-bladed horizontal-axis machine mounted on a steel substructure which anchors it to the sea floor. The asymmetric blades are 9 m long and may be pitched through 280° to optimise operation throughout the range of tidal current speeds, during both ebb and flow. Other design parameters include:

- Rated power: 500–2000 kW (site dependent)
- Rotor diameter: 21 m
- Nominal speed: 10 rpm

Figure 5.1 The HS300 tidal stream turbine (ANDRITZ HYDRO Hammerfest).

- Power regulation by variable pitch, speed and yaw
- Operating depth: 35–100 m
- Induction generator, output 500–2000 kW
- Nacelle weight: about 130 t
- Substructure weight: about 150 t (excluding ballast)
- Installation by heavy-lift vessel/barge, with remotely operated vehicle (ROV) support
- Lifetime/service: 25 years/every 5 years.

Winter weather conditions at EMEC during installation were extremely challenging but, within a few months, the device had been grid-connected, tested at maximum power, and was busy supplying electricity to houses and businesses on the Orkney island of Eday (see Figure 2.42).

Figure 5.2 HS1000 arrives in Orkney (ANDRITZ HYDRO Hammerfest).

During the rest of 2012 the device was put through a stringent series of performance and reliability checks in preparation for larger scale production. Heavily instrumented, its operation could be monitored from EMEC's facility on Eday, and also from the company's Glasgow office using mobile connections and an on-board camera.

At the same time plans were being finalised for an array of ten HS1000 tidal turbines in the Sound of Islay, off the west coast of Scotland between the islands of Islay and Jura (for the location of Islay, see Figure 1.11). Following two years of survey work on an environmental impact assessment (EIA), the 10 MW tidal power plant, a project with ScottishPower Renewables [3] as project developer, was consented in 2011 and scheduled for installation in 2014–2015. It is the first tidal turbine array to be approved by Marine Scotland, a directorate of the Scottish government responsible for managing Scottish waters. It has also received support from the Islay Energy Trust [4], a community-owned organisation that aims to develop renewable energy projects for the benefit of the local community. The site offers major advantages for a pre-commercial array: strong tidal flows in deep water and relative shelter from storms and waves.

Back in Section 1.4 we discussed ways in which information about turbine yield is given to the general public – and we used the Islay project as an example. As noted, the annual electricity production of the 10 MW array is expected to be about 30 GWh. This is similar to the entire needs of Islay, with its 3500 inhabitants (and 7 whisky distilleries!); alternatively, it is equivalent to the domestic needs of about 5000 Scottish households. We can see why a 10 MW array in the Sound of Islay makes good economic, as well as technical, sense; it is very close to a substantial island community which can use its electricity, and there is a suitable connection to the local grid.

Figure 5.3 shows the Sound of Islay, looking south from the village of Port Askaig with the Jura ferry approaching. The boat's course lies directly above the location of the turbine array. In case the tidal stream looks unconvincing in this photograph taken at slack water, we also include

Figure 5.3 Looking south along the Sound of Islay (Paul A. Lynn).

Figure 5.4 A vigorous tidal flow in the Sound of Islay (Paul A. Lynn).

Figure 5.4 – a view eastwards across the Sound towards Jura and its famous mountains, the Paps, which shows plenty of evidence of a tidal race in full flow. Its vigour is particularly noticeable at half-tide when tide and wind oppose one another.

As we have pointed out in Section 2.4.3, the siting of a turbine array on the sea bed is a far more complicated matter than might first appear. It is obviously important to place individual turbines in high peak and average stream velocities, to give the best possible electricity yield; but such issues as the nature of the sea bed, scouring by tidal streams and turbulence caused by the detailed bathymetry of the channel, must be taken into account. From an operational point of view it is important for the turbine to be easily accessible for maintenance and recovery, and the highest point reached by its blade tips at low tide must not risk interference with shipping. When siting a complete array, unwanted turbulence caused by one machine on its neighbours should be kept to a minimum while bearing in mind that turbines spaced unnecessarily far apart incur extra cabling costs. Careful investigation of all these factors for the Islay array produced the turbine layout scheme shown in Figure 5.5.

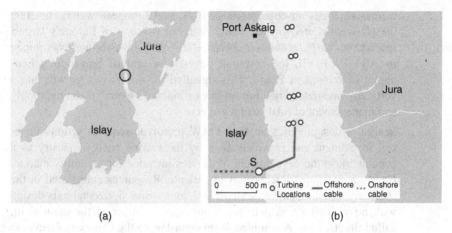

Figure 5.5 Geography of the Sound of Islay: (a) its location between Islay and Jura and (b) layout of the 10-turbine array.

ANDRITZ HYDRO Hammerfest and ScottishPower Renewables are confident that the HS1000 array will offset Islay's electricity requirements and that, in due course, larger arrays will be deployed in other locations, including the Pentland Firth off the north coast of Scotland.

5.2.2 Atlantis Resources

The story of Atlantis Resources Corporation [5] begins in Australia with the development of design concepts for tidal stream turbines. Scale model testing started in 2002, followed by tow-testing in ocean conditions. By 2006 the company had installed a 100 kW device and connected it to the Australian national grid in San Remo, on the Victorian coast some 50 km south of Melbourne. In the same year it established joint operations in Singapore, attracted by low costs, high quality engineering, impressive R&D capability and strong legal protection of intellectual property. In 2007 the company took the unprecedented step of tow-testing a 30 tonne device off the coast of Victoria without the prior use of scale models based on extensive computer simulation, and the following year saw a new 150 kW device installed and grid connected at San Remo. Design and development continued and a 500 kW machine was soon being tested in Singapore waters, where it displayed impressive 'water-to-wire' conversion efficiency.

The US investment bank Morgan Stanley became a shareholder in 2006 and Atlantis Resources acquired the bank's UK-based marine power business with the aim of offering turnkey projects from initial concept to final commissioning. A project office was opened in London with an increasing focus on opportunities in UK and European waters. In 2009 the Norwegian state-owned utility Statkraft [6], one of Europe's largest renewable energy generators, became an equity investor and was joined in 2011 by EDBI, the corporate investment arm of Singapore's Economic Development Board. This signalled the company's growth into a vertically-integrated organisation able to undertake overall planning, supply and management of tidal stream projects.

Scaling a design series up to a 1 MW prototype requires serious financial investment and is widely seen by the marine energy industry as a benchmark in the development of devices intended for the utility market. Nearly a decade of R&D effort by Atlantis Resources culminated in the company's first megawatt-scale device, an unusual horizontal-axis design with novel blades and twin rotors mounted at either of the main shaft, called the *AK1000*. Assembled at Invergordon on the Cromarty Firth near Inverness, it was transported by road and sea to EMEC [2] in Orkney in 2010. Unfortunately the blades soon revealed manufacturing faults and a strategic decision was made to revert to more traditional GRP (glass fibre reinforced plastic) blades and a single rotor. The modified design, unveiled in 2011 and named the *AR1000*, is based on proven and readily accessible components and is seen by the company as its flagship device for pre-commercial deployment.

Figure 5.6 shows the AR1000 loaded on a heavy transport vehicle for its journey to EMEC where it produced its first power for the Orkney grid in the summer of 2011. The three asymmetric blades are fixed to the hub and the turbine rotates to face the stream as the tide changes from ebb to flow. Main features and parameters include:

- Rotor diameter 18 m
- Rated power 1 MW at a tidal stream speed of $2.65\,\mathrm{m\,s^{-1}}$
- Rotor speed 12.6 rpm at rated power
- Gearbox ratio 6.15 : 1
- Permanent-magnet synchronous generator with 44 poles
- Turbine weight 130 t
- Sea-bed tripod foundation weight (with ballast) 1300 t.
- Fully-assembled height 22.5 m.

The AR1000 was shipped to Orkney on a dynamic positioning vessel equipped with a number of powerful thrusters, allowing it to hold station using GPS (see Figure 5.7). Fast, efficient, deployment and retrieval of a tidal turbine is imperative for making best use of limited operational time windows around slack tide, and for minimising installation costs.

Figure 5.6 The AR1000 begins its journey to EMEC (Atlantis Resources).

Figure 5.8 shows the device being lowered into the water at EMEC's Fall of Warness tidal test area. Atlantis Resources has put a lot of effort into the design of a special locating tool for extremely accurate alignment of the nacelle with the sea-bed foundation, allowing connection in no more than 90 min and retrieval in considerably less without any need for an underwater ROV or human divers. Following deployment at EMEC much valuable data was obtained on blade and system loadings, turbine control, grid connection, water tightness and marine coatings. The device was monitored and controlled from a cabin on the Orkney island of Eday (see Figure 2.42) containing all the power conditioning and control equipment.

Another important part of the AR1000's test programme was carried out at the UK's National Renewable Energy Centre (NAREC) [7] in Blyth, Northumberland. The centre offers advanced R&D and testing facilities for offshore wind, wave, tidal and electrical network technologies. Accelerated lifetime and extreme event testing is available for individual components, including gearboxes, generators and bearings, as well as for complete nacelles. A new 3 MW drive train test facility for wind and tidal turbines was opened in 2012 and one of its first customers was Atlantis Resources

231

Figure 5.7 At the quayside (Atlantis Resources; Mike Roper (photographer)).

Figure 5.8 The AR1000 is lowered into the sea at EMEC's tidal test site (Atlantis Resources).

and the AR1000 drive train. The company obtained performance data, including thermal properties of major electrical components under full electrical load, equivalent to four months of offshore testing in tidal ebbs and flows – a valuable time-saving for the company and no doubt for other developers in the future.

A principal aim of Atlantis Resources in the next few years is to provide turbines for the MeyGen project [8], a joint venture between Atlantis Resources, International Power Gdf Suez and Morgan Stanley. MeyGen Ltd was awarded rights in 2010 to develop one of the 'crown jewels' of the Pentland Firth – the Inner Sound located between Scotland's north coast and the island of Stroma – for which it holds a 25-year operational lease. The 3.3 km^2 site exhibits some of the fastest tidal currents in Scotland's impressive armoury – peaking at up to 5 m s^{-1} on spring tides – and has a potential installed capacity of 400 MW. MeyGen lodged plans for an initial 20 MW development in 2012 and is committed to establishing the Inner Sound as one of Scotland's pioneering projects in marine energy.

Atlantis also has projects under development in India and Canada. The Mundra Project in the state of Gujarat is Asia's largest planned tidal power project with a projected eventual capacity of 250 MW, and is due to start construction using Atlantis AR1000 series turbines. In 2011 Atlantis, along with consortium partners Lockheed Martin and Irving Shipbuilding, won a berth development right at the tidal energy demonstration facility operated by the Fundy Ocean Research Centre for Energy (FORCE) in Nova Scotia, Canada.

In 2012 Atlantis commenced a major research and development programme to reduce the cost of energy of tidal turbine arrays, partly funded by the Energy Technologies Institute (ETI) [9], a public–private partnership between the UK government and industry. The Tidal Energy Converter Demonstrator (TEC) project is being led by Atlantis in collaboration with engineering consultants Black & Veatch and systems integration specialists Lockheed Martin, as well as more than two dozen of the industry's leading developers and component manufacturers.

Traditionally, tidal energy developers have started from a given system architecture and then optimised their system around it, making incremental improvements with each turbine generation. Although this approach is cost effective in the short term, it is unlikely to lead to a truly optimised system configuration or offer the lowest cost of energy to compete effectively with other renewable technologies. In a radical new approach to R&D and cost reduction, the TEC project considers all current and potential components and configurations and optimises them to produce the lowest cost of energy for the entire system. Drawing on large volumes of information from

233

across the international tidal energy supply chain and using state-of-the-art modelling developed specifically for the project, the TEC programme led by Atlantis Resources requires multi-million pound upfront investment. However, it offers a rigorous approach to cost reduction and risk mitigation, aiming to drive down the cost of energy for the entire industry. Initial results have been encouraging and it is planned to deploy an optimised turbine system architecture in a representative offshore environment in the next few years.

5.2.3 Marine Current Turbines

The first major device designed and built by Marine Current Turbines (MCT) [10] was a 300 kW tidal turbine deployed in 2003 off the coast at Lynmouth in Devon, England. Intended as a temporary demonstration test facility, it delivered three years of invaluable operational data, followed by decommissioning and complete removal from site in 2009 – a world first for an offshore marine device. This wealth of experience was put to good use in the company's next design, known as SeaGen. By 2008 the Bristol-based company had deployed the 1.2 MW horizontal-axis device, the world's first utility-scale tidal stream turbine, in fast-flowing waters at the mouth of Strangford Lough in Northern Ireland (see Figure 5.9). In 2012 the company was bought by Siemens and is now operated as a business in the Ocean and Hydro division of the Energy Sector.

The company's approach to harnessing the energy of vigorous tidal streams is based on some important design features:

- Two-bladed rotors are 'parked' horizontally and raised above the sea surface for maintenance, avoiding the risks and difficulties of working underwater and the need for expensive support vessels.
- Two or more rotors and power trains are used per device, giving a measure of redundancy and useful options for power scale-up.
- When lowered, rotors are positioned towards the top of the water column where tidal flows are strongest.
- Advanced blade design techniques similar to those used for wind turbines produce fully-optimised, asymmetrical, blade profiles.
- Blades are pitch-controlled through 180° to give bidirectional operation, optimising energy capture on both ebb and flow tides without any need to yaw the device. Pitching is also used to 'spill' excess energy, and to allow rapid shut down from full power.
- All power conditioning equipment is housed within the support structure and the electricity transmitted to shore is fully grid-compliant.
- Great attention is paid to environmental impact.

Figure 5.9 Illustration of the SeaGen 1.2 MW device deployed in Strangford Lough (MCT/Siemens).

Marine Current Turbines offer two versions of the SeaGen technology: the 'surface-piercing' configuration shown in Figure 5.9, known as *SeaGen S*, with twin rotors and power trains supported on a crossbeam and a structure piled into the sea bed; and an underwater (non-surface-piercing) configuration known as *SeaGen U*, suitable for deeper water, with triple rotors supported by a structure hinged on the sea bed. We will cover the two versions in turn.

The commercial-scale prototype SeaGen installed in Strangford Lough was the first marine energy device in the UK to gain the status of a commercial power plant. It is suitable for maximum water depths of 35 m. Twin rotors give twice the energy capture for less than twice the cost, and the rotors may be independently operated. Each drives a gearbox and electrical generator, and the total rated power of 1.2 MW is delivered at all tidal stream speeds above 2.4 m s^{-1}. Further details include:

- Rotor diameter 16 m
- Rotor speed at rated power 14 rpm
- Gearbox ratio 69.8 : 1
- Gearbox type and weight: three-stage planetary, 12 t
- Weight of cross arm, rotors and power trains 120 t
- Electrical generation: two conventional induction generators supplying individual rectifiers and inverters.

The crossbeam is raised by a hydraulic jacking system, allowing the two-bladed rotors to be 'parked' for maintenance in a horizontal position raised above the sea surface, as shown in Figure 5.10. Another advantage of the crossbeam arrangement is that rotor performance is not degraded by the wake downstream of the pile; the crossbeam is streamlined and has a much less disruptive wake. The blades have many similarities with those used by today's wind turbines, with the important proviso that the much greater density of water produces very high loadings of up to 30 tonnes on relatively short blades. A lot of carbon fibre is therefore used in their construction (see Figure 5.11).

The consenting process for the device was very rigorous because Strangford Lough is a European-designated habitat with rich and varied flora and fauna, including grey and common seals. It is heavily protected by environmental laws including the stringent European Habitats Directive. Following extensive EIAs the company was initially permitted to operate the device under strict conditions. However a subsequent three-year monitoring programme revealed no significant impact on marine life and a full operating licence was granted. One important conclusion, for this and other tidal stream projects, is that although turbine rotors may look hazardous, seals and other marine life that can negotiate turbulent tidal streams are unlikely to collide with rotor blades moving with tip velocities of 15 m s^{-1} or less. Fortunately the rotors of Strangford Lough show every sign of being more attractive to tourists than seals!

When the device was first operated in 2008, the local grid proved too weak to accept the full rated output of 1.2 MW and the device was initially limited to 600 kW. Following grid enhancement it was run at full power and now regularly delivers in excess of 10 MWh per tidal cycle. The results of a lengthy test programme, reported in an excellent paper [11] published in 2010, illustrate many of the ideas we have discussed in Section 2.4.

Figure 5.10 The twin rotors are raised above the water for maintenance (MCT/Siemens).

The device's power curve shows output power increasing from about 50 kW at a cut-in speed of $1\,\mathrm{m\,s^{-1}}$ to the rated value of 1.2 MW above $2.4\,\mathrm{m\,s^{-1}}$, with slightly less generation on flood than ebb due to minor turbulence caused by the crossbeam. The rotor efficiency (C_p) on both tides varies between 45 and 52%, the latter corresponding to an impressive 88% of the Betz limit. Overall 'water to wire' system efficiency, taking account of losses in gearboxes, generators and power converters, is 40–45%. The device achieves a capacity factor well over 60%, with 95% availability, at this very energetic site. During spring tides in October 2012 it delivered 22.5 MWh in a single day, and by the end of 2012 had fed more than 6 GWh into the grid. Electricity yield is comparable to that of a well-sited 2 MW wind turbine but it is, of course, far more predictable.

MCT is currently bringing a more powerful version of its twin-rotor surface-piercing device, the *SeaGen S* 2 MW, to the market. Designed for commercial-scale projects, it has 20 m rotors and is rated at 2 MW. Its electrical and electronic systems allow a number of devices to be linked together in an array, minimising cabling requirements to shore. First deployments of the device are planned in 2015 at Kyle Rhea, Scotland and the Skerries, Wales (see below).

Figure 5.11 Construction of the SeaGen 1.2 MW rotor (MCT/Siemens).

Figure 5.12 The 3 MW SeaGen U configuration (MCT/Siemens).

We now turn to the *SeaGen U*, a 3 MW underwater (non-surface piercing) device being developed for deeper water sites. As Figure 5.12 shows, it consists of three 1 MW turbines deployed across the direction of the current, with the same extensively-tested rotors and power trains as the SeaGen S 2 MW but with a very different support structure. It is suitable for water depths exceeding 35 m and continues the company's design philosophy of raising turbines out of the water for maintenance, in this case using a streamlined structure hinged on the sea bed. The structure may be floated to the surface without any need for external lifting gear and carries power and instrumentation cables from the turbines to the sea bed.

Marine Current Turbines is involved in a number of projects in addition to the well-established power plant in Strangford Lough:

- *Kyle Rhea, Scotland.* This site is between the Isle of Skye and the Scottish mainland (see Figure 1.11). The company's plan is to deploy four SeaGen S 2 MW devices and is illustrated in Figure 5.13.
- *Skerries, Anglesey, North Wales.* This location is between Carmel Head and the Skerries Rocks off the northwest tip of the isle of Anglesey. The project, in collaboration with RWE-npower Renewables, is for five SeaGen S 2 MW turbines producing the equivalent of 20% of Anglesey's electricity demand.
- *Brough Ness, Orkney.* The company plans to install up to 50 SeaGen devices in the energetic tidal flows of the Pentland Firth, sited between the north-east tip of the Scottish mainland and South

Figure 5.13 Photomontage showing four SeaGen S 2 MW devices in Kyle Rhea, Scotland (MCT/Siemens).

Ronaldsay in the Orkney islands, with deployment planned for 2017–2020.

- *Bay of Fundy, Canada.* Marine Current Turbines has an agreement to supply a SeaGen U device to the Minas Pulp and Power Company as part of a facility operated by FORCE.

5.2.4 OpenHydro

The Irish company OpenHydro [12] is based in Greenore, Co Louth and has a manufacturing and research facility close to the port. The company was formed in 2005 following the purchase of world rights to the *Open-Centre* turbine technology pioneered over the previous decade in the United States by an Irish American, Herbert Williams [13]. The first large tidal stream device, rated at 250 kW, was deployed at EMEC [2] in 2006 and connected to the Scottish grid in 2008; by 2010 the company was testing a 1 MW full-scale prototype in the Bay of Fundy, Canada; and in 2012 the French naval contractor DCNS acquired a majority holding. Various international projects are currently in the planning pipeline and the company aims to support the sale of its turbines with turnkey installation and maintenance services.

The unique form of the Open-Centre turbine is illustrated in Figure 5.14. Designed for deployment on the sea bed, invisible to onlookers and

Figure 5.14 An Open-Centre tidal turbine for deployment on the sea bed (OpenHydro).

unaffected by shipping, its major components are a tripod support structure, a duct and a rotor with an open centre. There are some special features:

- The shaped duct improves the hydrodynamic efficiency of the rotor.
- A large-diameter permanent-magnet generator is contained within the duct.
- Marine life can pass safely through the open centre.
- A slow-moving rotor with shielded blade tips further reduces risks to marine life.

In 2006 OpenHydro became the first company to deploy a turbine at EMEC's Fall of Warness tidal test site (see Section 2.5.2). The scale and form of the rotor and duct may be seen in Figure 5.15 which shows the device during assembly. Rated at 250 kW, with a diameter of 6 m, it was the first tidal stream turbine in Scotland to deliver electricity to the national grid. Figure 5.16 shows it installed between the twin piles of a special test rig at EMEC. Two steel monopoles, grouted into sockets drilled into the sea

Figure 5.15 The 250 kW Open-Centre turbine (OpenHydro).

Figure 5.16 The turbine on its test rig at EMEC (OpenHydro).

bed, allow it to be raised and lowered for ease of maintenance. A suspended platform provides a safe working area and the rig allows modified versions of the device to be tested and optimised at minimum cost.

Utility-scale versions of the Open-Centre turbine are designed to operate on the sea bed and the company is steadily increasing power ratings as its R&D programme proceeds. A number of major European and North American projects are planned or in progress [12]:

- *Alderney, Channel Islands, UK*. The company is working with Alderney Renewable Energy [14] to install an array of Open-Centre turbines in the vigorous tidal waters surrounding the island of Alderney.
- *Bay of Fundy, Canada*. Collaborating with Nova Scotia Power [15] in a tidal demonstration project, OpenHydro deployed a 1 MW version of the turbine in the Minas Passage, Bay of Fundy, in 2009. The bay is famous for the largest tidal range in the world and the exceptional tidal flows in the Minas Passage, which can reach $5 \, \text{m s}^{-1}$ at spring tides, exerted forces on the turbine that exceeded its design ratings

and caused rotor blade failure. After recovery of the device from the sea bed in 2010 the project was put on hold pending detailed investigation and design modifications [12].

- *Brittany, France.* A major project to install tidal stream turbines off the coast of Brittany was initiated by Electricité de France (EDF) in 2004. The site chosen is the strait between the town of Paimpol and the small island of Bréhat, 120 km east of Brest, in water depths of about 35 m. In 2011 OpenHydro and EDF announced plans for an initial deployment of four 2.2 MW, 16 m diameter, Open-Centre turbines and the first unit was assembled at the DCNS shipyard in Brest and towed out to sea for a series of commissioning tests. When completed the Paimpol-Bréhat installation is expected to provide France with its first large-scale grid-connected tidal stream energy farm.

- *Pentland Firth, Scotland.* In 2010 OpenHydro and SSE Renewables won a licence to develop a tidal stream farm at the Cantick Head site in the Pentland Firth, with a potential installed capacity of 200 MW.

Figure 5.17 The installer barge (OpenHydro).

- *Torr Head, Northern Ireland*. In 2012 OpenHydro and Irish energy provider Bord Gáis Energy were awarded rights to develop a 100 MW tidal energy farm at Torr Head, County Antrim, off the north-eastern tip of Northern Ireland.

Successful deployment of large tidal stream turbines depends on fast and efficient procedures for accurate positioning of the devices above the sea bed, lowering them into the water and recovering them when necessary. Fast, oscillating, tidal races present an extremely difficult, even hazardous, working environment which would normally be avoided at all costs and full advantage must be taken of the very limited periods of slack tide. System designers recognise this as a key operational requirement and use considerable ingenuity to keep installation time and costs to a minimum. OpenHydro's solution is to use a custom-designed 'installer' barge to move their large turbines precisely into position (see Figure 5.17). The total load of a turbine and subsea base is typically around 1000 tonnes, with a height of 20–25 m. Figure 5.18, which shows a device loaded onto the installer prior to deployment, makes the point that time and tide wait for no man and that

Figure 5.18 Preparing for deployment at EMEC (OpenHydro).

Figure 5.19 A team effort off the coast of Orkney (EMEC).

night working may well be necessary. Finally, Figure 5.19 shows installer, turbine, tugs and support vessels in a carefully orchestrated manoeuvre to get everything on target.

5.2.5 Pulse Tidal

The tidal-stream device developed by Pulse Tidal [16], a company set up in 2007 in Sheffield, England, takes an unusual form. Instead of resembling an underwater version of a wind turbine, it uses pairs of horizontal hydrofoils that oscillate up and down across the stream as the tide ebbs and flows. The concept is illustrated by Figure 5.20, which envisages an array of devices mounted on the sea floor in fairly shallow water. Instead of 'sweeping' a circular area like a conventional turbine rotor, each hydrofoil taps energy from a rectangular area of water; by incorporating two pairs of hydrofoils in a single device the amount of energy captured is doubled.

Following an extensive R&D programme which included tank testing and theoretical modelling, the company deployed its first major demonstration device, the 100 kW *Pulse-Stream 100*, in 2009 near the mouth of the River Humber north-east of Sheffield. Designed for ease of access and mainte-nance, it comprises a piled structure supporting a single pair of hydrofoils, topped by a horizontal platform (see Figures 5.21 and 5.22). Power take-off is by mechanical levers, high-pressure hydraulics, a hydraulic motor

Figure 5.20 A photomontage showing an array of oscillating hydrofoil devices (Pulse Tidal Ltd).

Figure 5.21 Pulse-Stream 100 installed near the mouth of the River Humber (Pulse Tidal Ltd).

Figure 5.22 A view from above showing the twin hydrofoils raised out of the water (Pulse Tidal Ltd).

and a conventional electrical generator, similar to the system described in Section 3.2. The shallow water (9 m) and sheltered location in the Humber estuary made installation and operation relatively simple and the device successfully delivered electricity to the UK's national grid over a two-year period. Pulse-Stream 100 was decommissioned in 2012 but remained on location where it continued to attract visitors keen to see a pioneering example of tidal stream technology. The valuable experience gained by the company encouraged progress towards a much larger full-scale prototype.

Commercial versions of Pulse Tidal's device rated at 1 MW and above are designed to operate on the sea bed – as already shown in Figure 5.20. The company considers its approach offers a number of special features and advantages over horizontal-axis turbines:

- The device 'sweeps' a long low rectangle of water rather than a circle, making it particularly suitable for shallow water. This is illustrated by Figure 5.23.

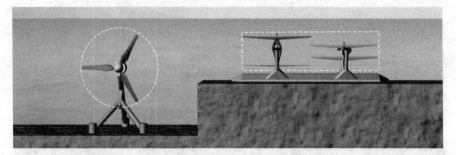

Figure 5.23 Pulse-Stream devices "sweep" shallow water more effectively than conventional turbines (Pulse Tidal Ltd).

- The length of the horizontal hydrofoils is not limited by the depth of water and allows commercial-scale devices to be deployed in depths of only 20 m.

- The nonrotating base is specially shaped to enhance flow over the hydrofoils.

- Shallow water is generally a less challenging near-shore environment, simplifying installation and maintenance and reducing the cost of grid connection.

- Survivability is enhanced by folding the blades down onto the base in rough conditions.

- Since the device is buoyant it can be fully assembled on shore and floated out for installation. Deployment and recovery are straightforward, requiring no special vessels (see Figure 5.24).

- The device is invisible in operation and poses no significant threats to wildlife.

The technical design of oscillating hydrofoils poses some unusual challenges, particularly in blade control. As in horizontal-axis turbines, the

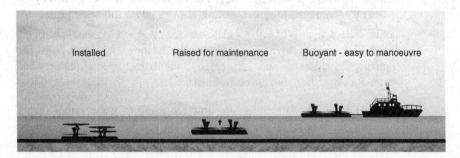

Installed   Raised for maintenance   Buoyant - easy to manoeuvre

Figure 5.24 The buoyant Pulse-Stream gives operational simplicity (Pulse Tidal Ltd).

blades are designed to generate lift forces at right angles to the incoming stream that translate into useful torque, while at the same time minimising unwanted drag. We have already discussed lift and drag in relation to turbine blades in some detail in Section 2.4.2.3, and many of the arguments are similar for oscillating hydrofoils. For example, as a blade moves and generates power, it creates its own stream which adds vectorially to the incoming tidal stream to produce a *resultant* stream that is actually 'seen' by the blade; and to be effective the blade must be held at an angle of attack – typically between 8° and 12° – to this resultant, generating plenty of lift while avoiding the stall condition. Achieving this goal with a horizontal blade is straightforward in two respects:

- The stream created by the blade is similar in strength along its whole length (whereas in a conventional turbine it increases linearly with distance from the hub). This means that the blade does not need to be twisted in order to preserve the angle of attack along its length, simplifying manufacture.

- Since the blade moves up and down across the tidal stream, generating power equally in both directions, its hydrofoil section is symmetrical.

However there are some complicating issues when it comes to blade control:

- As the blade moves up and down, describing an arc, it must be continuously pitched to maintain the desired angle of attack.

- The effective length of the blade's lever arm and, therefore, the torque produced by a given amount of lift, are both reduced as the blade moves away from its central position. This implies large variations in torque and power output as the blade goes through a complete cycle of oscillation.

- When the blade reaches the top or bottom of its travel, the angle of attack must be quickly reversed to generate lift (and movement) in the opposite direction.

Clearly, all this demands a sophisticated scheme of blade control to optimise energy capture. Another important consideration is blade speed and the time taken to complete a cycle of oscillation. If the blade moves too slowly, a lot of water passes through the swept area without doing useful work; but if it moves too rapidly, the high tip speed ratio reduces efficiency [13]. A fuller discussion of these and other issues may be found elsewhere [17].

Following the successful operation of Pulse-Stream 100 in the River Humber (Figure 5.25), the company announced its next major step along the road to commercialisation. Subject to local consultations and the necessary environmental assessments, it secured an agreement to lease an

Figure 5.25 Approaching Pulse-Stream 100 in the early morning (Pulse Tidal Ltd).

area of sea bed near Lynmouth in North Devon for deploying a 1.2 MW device. The site offers good tidal stream resources in fairly shallow water (18 m) close to shore, with a convenient grid connection, and is located in England's new South West Marine Energy Park. When deployed the commercial-scale device should help unlock the large tidal stream potential of the Bristol Channel.

5.2.6   Scotrenewables Tidal Power

Founded in the Orkney Islands in 2002, Scotrenewables Tidal Power [18] designed and built nine different scale models of its innovative floating tidal stream device over the following decade. The models, between 1/40th and 1/5th scale, were comprehensively tested in tanks and by towing at sea – see Figure 5.26 (you may also like to refer back to Figure 2.41). The company's first major prototype, a 250 kW device known as the *SR250*, was constructed at Harland and Wolff's shipyard in Belfast in 2010, from where it was taken to EMEC in Orkney and connected to the UK's national grid. During an

Figure 5.26 Testing the 1/5th scale model off the coast of Orkney (Scotrenewables Tidal Power Ltd).

extensive test programme in 2011–2012 the device proved highly stable at all stages of the tide, reached its nominal power rating and then surpassed it by generating a peak export power of 273 kW in strong spring tides. The invaluable technical and operational experience encouraged the company to press ahead with the design of its 2 MW commercial-scale prototype, the *SR2000*, supported by its three investor partners – the Fred Olsen Group, energy company TOTAL and ABB Technology Ventures.

The *SR250*, shown in Figure 5.27, was designed for ease of installation and recovery, operation and maintenance, and survivability in the harsh offshore environment – without any need for large and expensive support vessels. It consists of a floating cylindrical tube, 33 m long and 2.3 m in diameter, supporting twin horizontal-axis rotors splayed sideways at the stern (only one rotor is shown in the figure). Retractable rotor legs give the 100 tonne device two configurations: an operational mode with rotors down and the device secured to the sea bed by a compliant mooring system; and a transport/survival mode with rotors retracted. The design concept, which the company intends carrying forward to its commercial-scale devices, has a number of advantages:

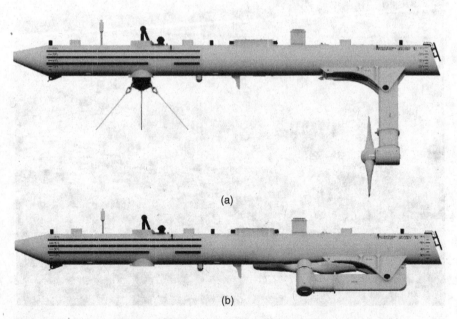

(a)

(b)

Figure 5.27 The two SR250 configurations: operational (a) and transport/survival (b) (Scotrenewables Tidal Power Ltd).

- The floating tube places the turbines in water regions with the highest tidal velocities, increasing energy availability compared with seabed mounted systems.

- A single-point mooring connection keeps the device headed accurately into the stream at all stages of the tide.

- The mooring system allows deployment in deep water.

- Most maintenance tasks can be carried out while the device is in the water.

- Connection and disconnection take less than 30 minutes.

- In storm conditions power generation ceases automatically, the rotor legs retract and the device assumes a streamlined shape with large waves passing smoothly over it.

The SR250 rotors have a diameter of 8 m and their counter-rotating fixed-pitch blades drive separate gearboxes and variable-speed electrical generators housed within subsurface nacelles. Power is routed into the main tube where it is converted to grid-compliant electricity by power conditioners before transforming up to 6.6 kV for export by subsea cable. Apart from the rotors and drive trains, almost all the device's equipment – control systems, hydraulics, transformers and switchgear – is contained within the tube.

Many tidal stream devices are mounted on sea bed substructures and must be raised for maintenance or repair. A floating device such as the SR250 is well suited to a catenary mooring system, rather as ships are moored offshore using cables and anchors. The innovative Scotrenewables system uses four catenaries fixed to the sea bed in a splayed pattern, brought up to the surface together and pulled into a 'mooring turret' on the device (see Figure 5.27a). Great attention has been paid to mechanical and electrical connection and disconnection, which must be fast, secure and reliable in the demanding environment of a vigorous tidal stream. One great advantage of single-point mooring with catenaries is that it allows the device to align itself passively with the tidal flow at all stages of the tide, much as a weather vane swings automatically to face the wind. Another advantage is the system's high degree of compliance (elasticity), which increases the device's survivability by 'cushioning' it against extreme sea conditions at the surface. Finally, the mooring system is cheap and easy to install compared with the massive structures often needed by sea bed devices.

The manufacture of large marine energy devices involves an impressive range of skills and techniques, and the SR250 is no exception. At the 'heavy' end of the scale, fabrication of the main tube involves cutting, rolling and welding steel plate to high tolerances; rotor blades demand exceptionally accurate forming and finishing. Intermediate in scale are the many subsystems – mechanical, hydraulic, electrical and electronic – that must be brought together to complete the device. Finally, effective monitoring and control to maximise electricity production without compromising safety and survivability call on advanced instrumentation and computer control. Some idea of the range of mechanical engineering needed to produce the finished product is given by Figures 5.28–5.30.

The SR250 was designed as a pre-commercial prototype to demonstrate the technology at large scale. Its size and power rating represent a prudent intermediate step between the one-fifth scale model in Figure 5.26 and a commercial-scale device. The extensive test programme at EMEC in 2011–2012 (Figure 5.31) put the SR250 through a carefully staged series of trials, allowing the company to assess technical performance and develop safe and efficient operating procedures. By the end of 2012 it had progressed from tow-trials using a modest-size work boat in gentle sea conditions to

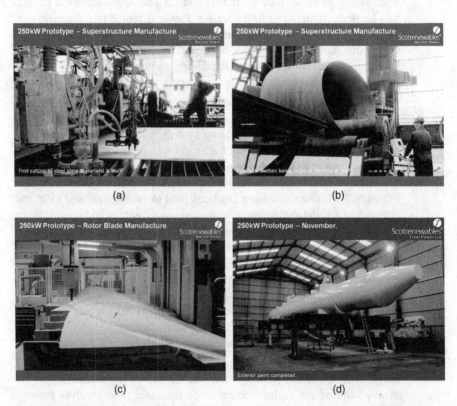

(a)

(b)

(c)

(d)

Figure 5.28 Manufacturing the SR250: (a) cutting steel plate, (b) rolling a tube section, (c) shaping a rotor blade and (d) finishing the steel tube (Scotrenewables Tidal Power Ltd).

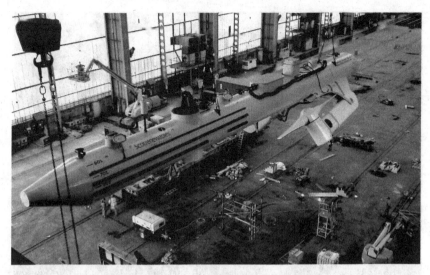

Figure 5.29 The SR250 nears completion (Scotrenewables Tidal Power Ltd).

Figure 5.30 The SR250's twin rotors (Scotrenewables Tidal Power Ltd).

Figure 5.31 The SR250 reaches Orkney waters (Scotrenewables Tidal Power Ltd).

full power generation at EMEC's Fall of Warness tidal test site, one of Europe's most challenging tidal stream environments (see Figure 5.32).

The company's latest device, known as the *SR2000*, is rated at 2 MW in a $3\,\mathrm{m\,s^{-1}}$ tidal stream and based on twin 1 MW turbines. The initial design used similar hydrodynamic and structural modelling techniques to those developed for the SR250, optimising the device for power capture, reliability and capital and operational costs over a 20-year life. Practical results from the SR250 test programme at EMEC provided data to verify the results of numerical models without scaling uncertainties, allowing reduced safety factors and more efficient design. The SR2000, with eight times the power rating of the SR250, is expected to have only four times its weight. Once again, great emphasis is being placed on straightforward installation, retrieval and redeployment using a modest and readily available workboat.

The SR2000 is intended for tidal array deployment – see Figure 5.33 – with intensive testing at EMEC prior to commercial demonstration and sales. In November 2012 the company was awarded an 'agreement for lease' by the UK's Crown Estate to develop a 30 MW tidal stream array in Lashy Sound, a site with peak spring rates of $3\,\mathrm{m\,s^{-1}}$ between the islands of Eday and Sanday in Orkney. After onsite resource assessments are completed and the necessary consents obtained, the project envisages an initial installed capacity of 10 MW with expansion to 30 MW by 2020.

Figure 5.32 The purpose-built SR250 monitoring and control suite at EMEC's Fall of Warness site, Orkney. The device can be seen through the right-hand window (Scotrenewables Tidal Power Ltd).

Figure 5.33 Visualisation of a large commercial array of Scotrenewables 2 MW tidal turbines (Scotrenewables Tidal Power Ltd).

257

5.2.7 Tidal Generation

Established in 2005 in Bristol, UK, Tidal Generation Limited (TGL) [19] became a wholly owned subsidiary of Rolls Royce, the global power systems company, in 2009. In 2013 Rolls Royce completed the sale of TGL to Alstom [20], the multi-national company which has ocean research and development activities based in Nantes, France.

In September 2010 TGL deployed its 500 kW prototype tidal turbine at EMEC's Fall of Warness tidal test site in Orkney, as part of the Deep-Gen III project co-funded by the UK's Technology Strategy Board. Operating in a water depth of 43 m, the device successfully fed over 250 MWh into the national grid, equivalent to the consumption of 300 households, and averaged 12 hours of generation per day. It was the first Scottish tidal project to receive UK Renewable Obligation Certificates (ROCs).

Figures 5.34–5.36 show the device before transportation to EMEC. The turbine rotor has three variable-pitch blades and is connected to a relatively standard drive train and power electronics. A separate steel foundation, in

(a)   (b)

Figure 5.34 The rotor and nacelle (a) and steel tripod foundation (b) of the TGL 500 kW device (Tidal Generation Ltd).

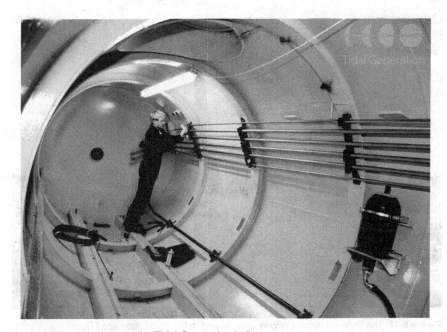

Figure 5.35 Inside the nacelle (Tidal Generation Ltd).

Figure 5.36 On the move to Orkney (Tidal Generation Ltd).

Figure 5.37 The buoyant nacelle allows easy towing to the installation site (Tidal Generation Ltd).

the form of a large tripod, is designed to secure the turbine to the sea bed. Since the nacelle is buoyant it may easily be towed to the foundation site (see Figure 5.37); then, using a ROV, the nacelle is winched down under water and secured with a patented clamping mechanism. When operating, the nacelle yaws at slack water to face the stream at an optimal angle. High voltage electrical and control connections are made automatically.

Tidal Generation installed a new 1 MW commercial-scale turbine at EMEC in early 2013. Mounted on the same tripod, it makes use of the valuable design and operational experience gained with the 500 kW device. Following a two-year test programme, the company plans the deployment of a demonstration array in UK waters as a prelude to commercial production.

Like its precursor, the 1 MW turbine may be easily floated to the installation site without large, expensive, support vessels and quickly secured to the seabed foundation during brief periods of slack tide. Once installed, the rotating nacelle aligns the rotor accurately to face the oncoming stream at all stages of the tide, making best use of its high-efficiency variable-pitch blades. This approach optimises energy capture when – as is often the case – the stream direction does not change by exactly 180 ° between flow and ebb phases. The ability to yaw the nacelle and pitch the blades also adds greatly to survivability in extreme conditions, especially when used in conjunction with sophisticated computer control. A relatively standard drive train, similar to those used over many years in large wind turbines (see Section 3.4.3.1), avoids the risks associated with 'prototypes within prototypes' and enhances reliability.

In more detail, technical characteristics of the 1 MW device include:

- Rotor diameter 18 m
- Rotor speed at rated power 14 rpm
- Rated power of 1 MW achieved at a water speed of 2.7 m s^{-1}
- Cut-in water speed 1 m s^{-1}
- Maximum operating water speed 3.4 m s^{-1}
- Drive train: epicyclic gearbox, induction generator, electronic power converter.
- Turbine length 21 m, height 5 m, weight 135 t
- Installed water depth 35–80 m
- Power export via subsea cable at 6.6 kV
- Rapid deployment/retrieval in 20 min
- Autonomous operation and remote control
- Modular construction for ease of maintenance and repair
- Planned service life 30 years.

Tidal Generation is operating the 1 MW turbine in a project partnership with the ETI [9], whose marine energy programme focusses on the industry's main technical challenges by supporting sea trials of pre-commercial devices and encouraging the development and demonstration of key technologies, systems and tools. The consortium project with Tidal Generation and a number of other developers and institutions is known as ReDAPT (Reliable Data Acquisition Platform for Tidal) and aims to collect and publish significant data on tidal stream energy systems for the benefit of the industry as a whole. Topics covered include advanced computer modelling of fluid flows and turbine arrays; instrumentation, control and operational procedures; and environmental impacts of tidal turbines. The Tidal Generation 1 MW turbine at EMEC makes a key contribution to the programme and is being tested in various operational conditions over a two year period. Detailed environmental and tidal stream information is helping the marine renewable industry understand the challenges to be addressed as tidal technology is developed on a commercial scale. ETI hopes to see installed capacity of marine energy devices reach 2 GW by 2020, moving towards 30 GW by 2050 with electricity costs competitive with more conventional renewables, including offshore wind. All this adds to public and industry confidence in tidal turbine technologies, which depends on providing comprehensive environmental impact and performance data as well as demonstrating reliable new turbine designs.

References

1. Andritz Hydro Hammerfest, www.hammerfeststrom.com (accessed 24 April 2013).
2. EMEC, www.emec.org.uk (accessed 24 April 2013).
3. ScottishPower Renewables, www.scottishpowerrenewables.com (accessed 19 July 2013).
4. Islay Energy Trust, www.islayenergytrust.org.uk (accessed 19 July 2013).
5. Atlantis Resources Corporation, www.atlantisresourcescorporation.com (accessed 24 April 2013).
6. Statkraft, www.statkraft.com (accessed 24 April 2013).
7. NAREC, www.narec.co.uk (accessed 24 April 2013).
8. MEYGEN, www.meygen.com (accessed 24 April 2013).
9. ETI, www.eti.co.uk (accessed 24 April 2013).
10. Marine Current Turbines Ltd, www.marineturbines.com (accessed 24 April 2013).
11. P.L. Fraenkel, *Development and Testing of Marine Current Turbines SeaGen 1.2 MW Tidal Stream Turbine*. 3rd International Conference on Ocean Energy, Bilbao, Spain (2010).
12. OpenHydro, www.openhydro.com (accessed 24 April 2013).
13. J. Hardisty. *The Analysis of Tidal Stream Power*, Wiley-Blackwell: Chichester (2009).
14. Alderney Renewable Energy, www.are.gb.com (accessed 19 July 2013).
15. Nova Scotia Power, www.nspower.ca (accessed 24 April 2013).
16. Pulse Tidal Ltd, www.pulsetidal.com (accessed 24 April 2013).
17. www.esru.strath.ac.uk/EandE/Web_sites/05-06/marine_renewables/technology/oschydro.htm (accessed 19 July 2013).
18. Scotrenewables Tidal Power Ltd, www.scotrenewables.com (accessed 24 April 2013).
19. Tidal Generation Ltd, www.tidalgeneration.co.uk (accessed 24 April 2013).
20. ALSTOM, www.alstom.com (accessed 24 April 2013).

Index

Printed in the United States
By Bookmasters